Springer Theses

Recognizing Outstanding Ph.D. Research

Aims and Scope

The series "Springer Theses" brings together a selection of the very best Ph.D. theses from around the world and across the physical sciences. Nominated and endorsed by two recognized specialists, each published volume has been selected for its scientific excellence and the high impact of its contents for the pertinent field of research. For greater accessibility to non-specialists, the published versions include an extended introduction, as well as a foreword by the student's supervisor explaining the special relevance of the work for the field. As a whole, the series will provide a valuable resource both for newcomers to the research fields described, and for other scientists seeking detailed background information on special questions. Finally, it provides an accredited documentation of the valuable contributions made by today's younger generation of scientists.

Theses may be nominated for publication in this series by heads of department at internationally leading universities or institutes and should fulfill all of the following criteria

- They must be written in good English.
- The topic should fall within the confines of Chemistry, Physics, Earth Sciences, Engineering and related interdisciplinary fields such as Materials, Nanoscience, Chemical Engineering, Complex Systems and Biophysics.
- The work reported in the thesis must represent a significant scientific advance.
- If the thesis includes previously published material, permission to reproduce this must be gained from the respective copyright holder (a maximum 30% of the thesis should be a verbatim reproduction from the author's previous publications).
- They must have been examined and passed during the 12 months prior to nomination.
- Each thesis should include a foreword by the supervisor outlining the significance of its content.
- The theses should have a clearly defined structure including an introduction accessible to new PhD students and scientists not expert in the relevant field.

Indexed by zbMATH.

More information about this series at http://www.springer.com/series/8790

Evan Shockley

Study of Excess Electronic Recoil Events in XENON1T

Doctoral Thesis accepted by The University of Chicago, USA

Springer

Evan Shockley
University of California San Diego
La Jolla, CA, USA

ISSN 2190-5053 ISSN 2190-5061 (electronic)
Springer Theses
ISBN 978-3-030-87751-4 ISBN 978-3-030-87752-1 (eBook)
https://doi.org/10.1007/978-3-030-87752-1

This Springer imprint is published by the registered company Springer Nature Switzerland AG
The registered company address is: Gewerbestrasse 11, 6330 Cham, Switzerland

Supervisor's Foreword

I am delighted to introduce this work by my former graduate student, Dr. Evan Shockley. Evan, as a member of the XENON collaboration, had the opportunity to search for potential signs of new physics by studying electron-mediated interactions in the core of the XENON1T Time Projection Chamber. The multi-tonne xenon detector was operated steadily from 2016 to 2018 and, featuring throughout its operation an ultra-low background, allowed us to reach an unprecedented sensitivity to dark matter interactions and other physics beyond the standard model.

Evan's work was focused on electron recoils with energies near the detector threshold. This energy range, roughly from 1 to 210 keV, is extremely appealing due to its sensitivity to potential new physics. At the same time, this regime, with technologies pushed to their edges, is also more prone to be affected by new instrumental effects and potential new sources of background not yet observed or accounted for in previous analyses. During this analysis, Evan stumbled upon, to our surprise, an unexpected excess of interactions, hardly explained by a statistical fluctuation of the sole background when all known sources were accounted for. He sat on these data for almost an entire year, improving the background models and testing a large variety of effects that could have produced the excess, at the level of both event reconstruction and detector response characterization. The more his tools got refined, the more the significance of the excess with respect to known backgrounds increased. Once ruled out all known effects, and with a solid and exciting excess in our hands, Evan dived into its possible interpretations.

As Evan points out in his thesis, among all *mundane* explanations of the excess the only one that we could not rule out with high confidence was the presence of a tiny contamination of tritium mixed with the xenon, at the level of 3 atoms of tritium per kg of xenon. While we took several precautions during operations to minimize the amount of tritium that could have reached the detector, the uncertainties on the effectiveness of some suppression techniques did not allow us to fully rule out this possibility. Decaying tritium atoms would emit low-energy electrons with an energy spectrum compatible, within statistical fluctuations, to the one of the excess. Nevertheless, the statistical analysis of the excess energy spectrum showed that it would be much better described by processes, other than the tritium decays, involving new physics. While the range of models involving a new particle interacting within the xenon target via electron recoils is extremely broad, the XENON collaboration considered three alternative hypotheses that would induce in the detector a signature similar to the one observed. Evan developed models for solar axions, solar neutrinos with anomalous magnetic moment, and dark photons and, comparing with the data, he drew conclusions on the value of several basic parameters involved in the signal models. Some of these interpretations turned out to be in tension with other respective indirect measurements, and some are exploring a completely new territory. Part of Evan's studies ended up in a collaboration paper [*E. Aprile et al., Phys. Rev. D 102, 072004 (2020)*]. The work contained in his thesis goes anyway beyond what is presented in the paper, and in the last two chapters, Evan shows original work meant to further strengthening the significance of the effect.

Evan's journey throughout this analysis has been extremely exciting for us but always guided by a strong scientific ethic, with the goal of sharing with the community the observation of an excess while recognizing the incapability of XENON1T data in discriminating between the more mundane and the more exciting hypotheses. The XENONnT detector, the successor of XENON1T, is being operated right now and its data will help us in figuring out if Evan had stumbled upon a tiny new and unexpected background or on something much more transformative.

University of Chicago
Chicago, IL, USA

Luca Grandi

Acknowledgments

I am deeply grateful for my advisor, Prof. Luca Grandi, for his guidance, generosity, and patience. I am also thankful for Dr. Richard Saldanha for being an invaluable mentor early in my graduate career, and similarly Dr. Jacques Pienaar for being a great mentor and friend. Jingqiang Ye and Dr. Michelle Galloway were a pleasure to work with on the electronic recoil analysis. I would like to thank the Physics Department at the University of Chicago and the Kavli Institute for Cosmological Physics for their generous support. Lastly, I want to thank my wife, Stephanie, and all the friends I have made at the University of Chicago for making the past 6 years ones I will always cherish.

Contents

1 Searching for New Physics with XENON1T 1
1.1 The XENON1T Detector 2
1.1.1 The S1 Signal 4
1.1.2 The S2 Signal 5
1.1.3 Position Reconstruction 6
1.1.4 Particle Discrimination: Electronic vs. Nuclear Recoils 6
1.1.5 Background Reduction 8
1.2 Physics Reach of XENON1T 8
1.2.1 Solar Axions 9
1.2.2 Neutrino Magnetic Moment 15
1.2.3 Bosonic Dark Matter 17

2 Event Reconstruction and Selection 19
2.1 Energy Reconstruction and Resolution 19
2.1.1 Energy Reconstruction 19
2.1.2 Energy Resolution 27
2.2 Detection Efficiency 29
2.2.1 Single Photoelectron Acceptance 30
2.2.2 Reconstruction Efficiency 34
2.2.3 Efficiency as Function of Charge 39
2.2.4 Efficiency as Function of Reconstructed Energy 41
2.3 Event Selection 44

3 Background + Signal Modeling and Statistical Methods 47
3.1 Background Model 47
3.2 Signal Modeling 57
3.3 Statistical Framework 59

4 Results 67
4.1 Background-Only Fit 67
4.2 Investigating the Excess 71

4.3 Potential Backgrounds ... 78
4.3.1 Xenon X-Rays ... 78
4.3.2 ^{37}Ar ... 78
4.3.3 Tritium ... 80
4.4 Searches for New Physics in XENON1T ER Data ... 83
4.5 Search for Solar Axions ... 84
4.5.1 Main Results ... 84
4.5.2 Interpretation Under Specific QCD Models ... 89
4.5.3 Including Tritium as a Background ... 93
4.6 Search for an Enhanced Neutrino Magnetic Moment ... 95
4.6.1 Main Result ... 95
4.6.2 Including Tritium as a Background ... 96
4.7 Bosonic Dark Matter Results ... 97
4.8 Science Run 2 ... 100
4.9 Time Dependence ... 102
4.10 Updated Analysis ... 103
4.10.1 Summary of Updates ... 103
4.10.2 Results of Updated Analysis ... 105

5 Conclusions and Outlook ... 109
5.1 Summary ... 109
5.2 Outlook: XENONnT ... 111
5.3 Conclusion ... 112

References ... 113

Chapter 1
Searching for New Physics with XENON1T

Abstract Dark matter is one of biggest open questions in physics today. In attempts to answer this question, many ultra-sensitive particle detectors have been developed to search for dark matter interactions here on Earth. In this chapter, we introduce the question of dark matter and describe in detail one such detector that searches for it: XENON1T. While XENON1T was designed to primarily search for the WIMP, a particularly well-motivated dark matter candidate, we explain that its low background level, low energy threshold, and large exposure allow us to use XENON1T to search for other signals physics beyond the Standard Model, including other dark matter candidates. This thesis focuses on the search of three such signals using XENON1T: solar axions, an enhanced neutrino magnetic moment, and bosonic dark matter.

There is overwhelming evidence for the existence of dark matter [60], a substance whose gravitational effects on cosmological and astrophysical scales are observed regularly, but has never been detected directly in a laboratory setting. With all evidence pointing to dark matter outnumbering "normal" matter—every atom, molecule, organism, planet, star, galaxy, etc.—by a factor of $\sim$5, the nature of dark matter is one of the biggest open questions in physics today. What is it, and how does it interact with normal matter? Answers to these simple questions would have deep and far-reaching impacts on humanity's understanding of our Universe on a fundamental level.

It is generally accepted that dark matter is likely composed of one or more species elementary particles not described by the Standard Model of Particle Physics, abbreviated the Standard Model or the SM. We have evidence that essentially every galaxy, including our own Milky Way, sits within a dark matter halo, meaning that our Solar System (and thus the Earth) is likely propagating constantly through a field of dark matter while it orbits the center of the galaxy. If dark matter is indeed composed of a new particle and if that particle does (rarely) interact with normal matter, as many theoretical candidates predict, then in principle a very sensitive particle detector should be able to detect this dark matter directly. Over the last

E. Shockley, *Study of Excess Electronic Recoil Events in XENON1T*,
Springer Theses, https://doi.org/10.1007/978-3-030-87752-1_1

several decades many experiments have been carried out to do just that, but no convincing dark matter signal has been observed to date [98, 147].

The XENON collaboration is a leader in the dark matter direct detection field that has designed and operated a family of two-phase xenon detectors that were primarily designed to search for a particularly well-motivated dark matter candidate, the Weakly Interacting Massive Particle (WIMP). The XENON1T experiment, the most recent effort of the XENON collaboration, has recently published the most stringent constraints on WIMP interactions [31, 33], but observed no signal of the elusive dark matter. In their efforts to detect WIMP dark matter, XENON1T and similar experiments have now achieved sensitivities to several other types of *new physics*, including other dark matter candidates as well as new interactions of known particles.

In this thesis, we use XENON1T data to search for (1) solar axions, (2) an enhanced neutrino magnetic dipole moment, and (3) bosonic dark matter. Similarly to that of WIMP dark matter, a detection of any of these three signals would revolutionize our understanding of the Universe in a deep and fundamental way. A secondary goal of this thesis, in addition to the search results themselves, is to highlight the sensitivity of XENON1T to physics outside the WIMP paradigm and in doing so motivate a broader physics program in future dark matter experiments.

1.1 The XENON1T Detector

XENON1T is a dual-phase (liquid/gas) xenon time projection chamber (LXe TPC) that operated underground at Laboratory Nazionali del Gran Sasso (LNGS) in central Italy from 2016–2018. With a total mass of 3.2 metric tons (t), it is the largest dual-phase TPC deployed as of the time of writing (mid-2020).

A schematic of the XENON1T detector is shown in Fig. 1.1. The cylindrical TPC, instrumented with photomultiplier tubes (PMT) arrays lining both the top and bottom of the detector, is used to search for light produced by particle interactions within the 2-t active volume of liquid xenon. A thin layer of gaseous xenon separates the top PMT array and the liquid active volume, and the walls of the TPC are made of highly reflective PTFE to maximize light collection in the PMTs. An additional ton of liquid xenon surrounds the TPC to serve as detector shielding from outside radiation.

When radiation—be it well-known particles such as photons (γ-rays and/or X-rays), electrons (βs), neutrons (n), and neutrinos (ν), or a new type of particle such as dark matter—interacts in the XENON1T detector, energy and momentum are transferred to the xenon, resulting in an *electronic recoil* if the interaction was with the atomic electrons, or in a *nuclear recoil* if the interaction was with the xenon nucleus. Both types of interactions produce a combination of scintillation light (photons) and ionization charge (electrons) that can be detected with a LXe TPC.

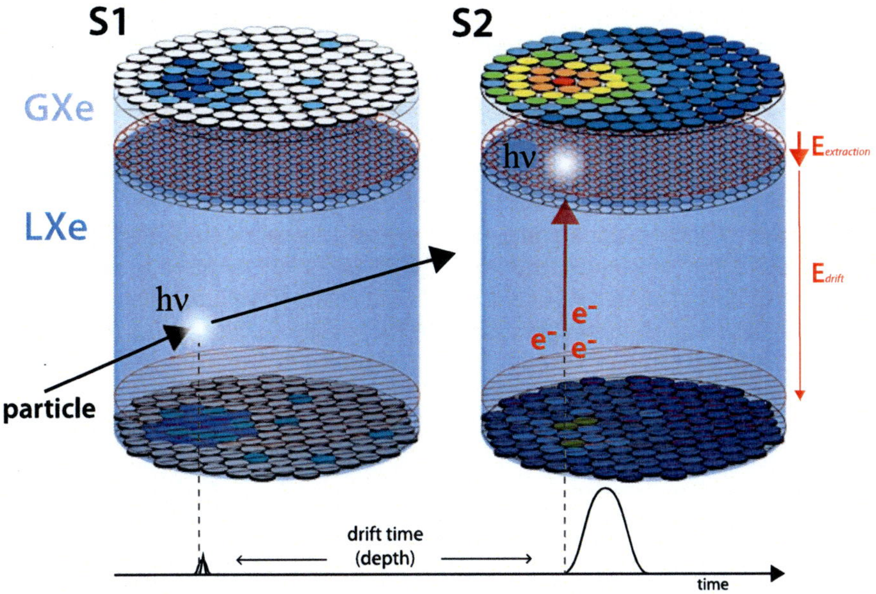

Fig. 1.1 Schematic showing the XENON1T detector and detection principle. Credit: Lutz Althüser

Xenon scintillates with a wavelength of 178 nm, which in purified liquid xenon has an absorption of $\simeq$50 m [30]. This means that liquid xenon is essentially transparent to its own scintillation light, and thus photons produced by interactions in the center of the liquid active volume can reach the PMTs that line the top and bottom of the detector. A scintillation signal produced in the liquid phase and detected by the PMTs is referred to as an S1.

Electron-ion pairs produced in a recoil would rapidly recombine in absence of an electric field; thus, to prevent this, XENON1T has a uniform electric field applied across the active volume that drifts the ionization electrons away from the interaction point and towards liquid-gas interface at the top of the detector.[1] At this interface ionization electrons are extracted into the gaseous phase and further accelerated, producing another light signal via electroluminescence [23]. This second light signal, proportional to the number of extracted electrons and captured by the same PMTs, allows for a measurement of the ionization charge produced in an interaction. We refer to this signal as an S2.

The detection of the S1 and S2 signals allow us to reconstruct several features of an interaction within the liquid xenon. The total amount of charge collected by the PMTs is related to the total energy deposited in the interaction. Since the uniform electric field drifts the ionization charge at a constant average velocity, the time

[1] The positively charged ions are accordingly drifted to the bottom of the detector.

difference between the S1 and S2 determine the depth or z position of the event. The x and y positions can be reconstructed from S2 signals observed in the top PMT array, thus allowing for full 3D position reconstruction. Further, the ratio of the S1 and S2 signals allow us to differentiate between different types of recoils, namely the electronic- and nuclear-recoils mentioned above. Since different particles tend to induce different types of recoils, this allows for a form of a particle identification (see Sect. 1.1.4).

Combining the detection principles described above with an unprecedented low background level and large target mass, XENON1T can be used to search for a number of ultra rare particle interactions, such as those from dark matter. A large detector mass scales up the probability of dark matter interacting in the detector, and a lower background rate means a smaller signal would be observable over background.

For detailed reviews of LXe TPC technology, we refer to [23, 70]. Some of the main points are briefly highlighted below.

1.1.1 The S1 Signal

The S1 signal arises from prompt scintillation of the liquid xenon medium. On a microphysics level, this scintillation comes from the relaxation of excited xenon dimers, called excimers. These excimers are produced via scatters of radiation in the xenon medium as well as *recombination*, in which xenon ions produced in the scatter event are able to snatch up free electrons before they escape. Excimers in either singlet or triplet states are possible, where the singlet/triplet fraction depends on the density of ionization. This ionization density is in turn related to details of the event that occurred: the energy deposited, the nature of the primary particle, and whether it interacted with the atomic electrons or the nucleus. Regardless of a singlet/triplet state, these excimers decay to the ground state via release of a characteristic VUV photon of 178 nm.

For some detectors (liquid argon, for example) the decay times of the singlet and triplet states are quite different, allowing for particle discrimination based on the time profile of a scintillation signal; however, in LXe this difference between the singlet (3 ns) and triplet (24 ns) is quite small, especially when compared to the typical collection time of S1 photons of $O(100)$ ns and the digitizer sample width of 10 ns. Thus *pulse shape discrimination* is not currently pursued in XENON1T.

Scintillation photons can travel long distances in pure liquid xenon but are easily absorbed by impurities, especially water [23]. In order to maintain the required purity level for detection of these photons, the xenon gas is constantly purified through hot zirconium getters, suppressing the water concentration to $\lesssim$1 ppb in XENON1T. The high level of purity allows for a high light yield and an analysis threshold as low as 1 keV for electron-mediated interactions.

Photons that reach the photocathodes of the PMTs can then produce photo-electrons (with quantum efficiency $\sim$30–40% [51]), which are then amplified by the dynode chain inside the PMTs [105]. The total charge collected by the PMTs is proportional to the number of detected photons in the S1, with resolution that depends on the PMT response. This charge is expressed in units of photoelectrons (PE), each PMT having been normalized according to its mean charge of a single photoelectron (SPE), determined via calibration.

1.1.2 The S2 Signal

The S2 signal originates from ionization electrons produced in an interaction that are drifted and extracted into the gas phase. Some of the original ionization is lost to recombination (i.e. converted to the scintillation signal), with the total amount depending on the density of the ionization track and the applied electric field. The electrons that survive recombination are then drifted to the top of the detector with a constant average velocity, where they are extracted into the gas via a strong electric field applied at the liquid-gas interface. However, some charge can be lost during the drift due to electronegative impurities in the xenon, especially O_2. For a concentration of ionization electrons c_e, the probability of an electron being lost to an impurity within a drift distance dz scales with the concentration of impurities c_I as well as c_e and z themselves. That is,

$$dc_e \propto -c_e \cdot c_I \, dz. \tag{1.1}$$

Introducing a proportionality constant k and solving the above differential equation, we have

$$c_e(z) = c_e^0 \cdot e^{-k\,z}, \tag{1.2}$$

where c_e^0 is the initial concentration of ionization charge. In order for this charge to be detected, it has to be drifted and extracted into the gas layer the top of the TPC; hence, for the same number of electrons produced, S2s near the bottom of the detector will be smaller than those at the top.

Since the electron cloud drifts with a constant velocity, the position dependent effect described above can easily be converted to a temporal effect, whereby the amount of charge lost depends on how long the electron cloud has been drifting. After the electron cloud has been drifted to the top of the TPC, resulting in a detected S2, the total drift time t_d can be identified as the time difference between the S1 and S2 signals. The amount of charge that reaches the liquid-gas interface can therefore be written

$$c_e\,(t_d) = c_e^0\, e^{-t_d/\tau}, \tag{1.3}$$

where on account of the variable change we define a new constant τ. This parameter, called the *electron lifetime*, is a measured value that encapsulates the xenon purity.

A large electron lifetime, ideally in excess of the maximum drift time of the detector, ensures minimal suppression of S2 signals due to charge loss. As shown in Sect. 2.1.1, (1.3) can be inverted to correct S2 signals for this suppression effect.

Most, but not all, of the charge that reaches the liquid-gas interface is then extracted into the gas phase, with the fraction given by the *extraction efficiency*. Due to local differences in the extraction field this efficiency varies slightly with x and y, but the average value measured to be 96% in XENON1T [37]. After extraction, the electrons are accelerated to the anode, resulting in proportional scintillation through a similar excimer mechanism as in the S1 case (though in the gas phase) and detected by the same PMTs. In XENON1T, a single electron results in an average S2 signal of ~28 PE, a quantity referred to as the *single electron gain* [37].

S2s have a time profile of $O(\mu s)$, in contrast to S1s with $O(ns)$. Further, this time structure has a strong spatial dependence in the detector, with deeper S2s having larger widths due to longitudinal diffusion during the electron drift [1]. These strong differences in pulse shape allow for straightforward discrimination between S1 and S2 signals, which is done using algorithms in the reconstruction software [2].

1.1.3 Position Reconstruction

The S1 and S2 signals can be used to reconstruct the position of an event in three dimensions. The z or the depth is determined from the drift velocity and the time difference between an S1–S2 pair. The x and y coordinates (or r and θ in cylindrical coordinates) are determined from the amount of light seen in specific PMTs in the top array (see Fig. 1.1). This is done using two algorithms in XENON1T: (1) a neural net algorithm trained using the Fast Artificial Neural Network Library (FANN) and (2) a likelihood maximization of the S2 pattern in the top array, referred to as Top Pattern Fit (TPF). The FANN is the main algorithm used, with the TPF algorithm primarily a cross-check. Both were trained with Monte Carlo (MC) simulated data using Geant4 [6] to propagate photons throughout the full detector geometry.

1.1.4 Particle Discrimination: Electronic vs. Nuclear Recoils

A major advantage of LXe TPCs is the ability to discriminate different types of interactions in the detector, especially *electronic recoils* (ERs) and *nuclear recoils* (NRs). ERs, in which the primary ionizing particle interacts with the atomic electrons, are induced by γs, X-rays, and βs, whereas NRs, in which the interaction is with the nucleus, are produced from, e.g., neutrons. Neutrinos can induce both ERs and NRs, but are not expected to contribute much in XENON1T due to their small cross sections.

Discrimination of ER and NR events is also performed with the S1 and S2 signals. NR events produce a denser track of charge in their wake, which leads to a

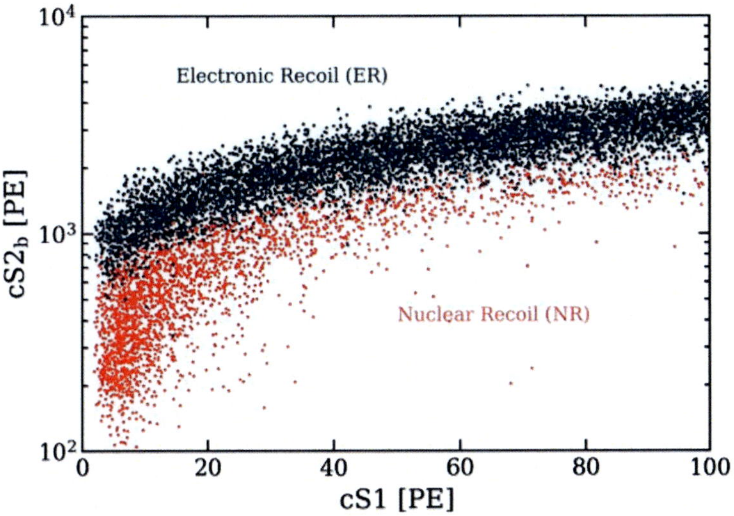

Fig. 1.2 Comparison of ER and NR calibration data in the space of S2 vs S1 to illustrate discrimination capability. The axes are corrected (hence the *c* prefix) for spatial effects in the TPC, and only the bottom PMT array is used for the S2 signal (hence the 'b' subscript)

larger fraction of recombination than in the ER case. This means that, for a signal with equal number of quanta produced, an NR would display a larger (smaller) number of photons (electrons) than an ER. Thus, ER and NR can be discriminated in the space of S2 vs. S1, as shown in Fig. 1.2.

For the sake of clarity it is worth stressing that the information above refers to the primary *ionizing* particle, not necessarily to the particle that triggered the event in the first place. For example, a γ Compton-scattering on a xenon atom induces the emission of an atomic electron. This electron then ionizes and excites the surrounding xenon atoms, resulting in the detection of S1 and S2 signals.

Additionally, it is noteworthy that NR signals are "quenched" due to additional energy loss to heat, which is not detected in LXe TPCs. This implies that the energy scales of ER and NR signals are different, as the light/charge yield for a given energy is smaller for NRs than for ERs. For more discussion on the differences between ER and NR, see [36, 151].

WIMP dark matter is expected to interact only with a xenon nucleus, thus inducing NRs, while the dominant backgrounds, βs and γs yield ERs. Thus, this NR/ER discrimination allows for a strong suppression of background in the NR detection channel and is a major factor in XENON1T achieving world-leading sensitivity to WIMP dark matter [31]. However, as shown in Fig. 1.2, this discrimination is not 100% effective; there is some overlap in the populations of ER and NR events. For this reason, in order to continue to reach better sensitivity to WIMPs, the ER background in XENON1T has also been suppressed to unprecedented levels. This ultra-low ER background makes XENON1T sensitive to interesting physics in the *ER channel*, as well as the historically prioritized NR channel. These ER signals are the focus of this thesis, as we describe in more detail below.

1.1.5 Background Reduction

XENON1T is designed to search for low-energy, keV-scale signals. There are innumerable backgrounds at these energies from mundane background radiation of common isotopes as well backgrounds from cosmic rays, making it difficult to search for ultra rare interactions in the same energy range. Thus one of the largest challenges (and successes) of XENON1T is the suppression of these backgrounds by many orders of magnitude, using a number of innovative methods.

First, operating underground at LNGS, the detector is shielded from cosmic rays by 1400 m of rock above it. The TPC itself is shielded again by a 10 m tall water tank, instrumented with 8″ PMTs, which acts as both passive shielding as well as an active muon veto. Every component in the detector was carefully screened, with low-radioactivity materials used wherever possible. Liquid xenon effectively shields itself, where most radiation originating outside the detector interacts within a few cm from the edge due to the medium's high Z and high density. Combining this with 3D position reconstruction allows for us to define a *fiducial volume* in the analysis, where we only consider the inner region of the detector to further reduce background. Background sources that cannot be removed by such shielding methods, i.e. radioactive krypton and radon isotopes in xenon itself, were reduced using cryogenic distillation [28]. Lastly, xenon is itself radiopure, with very few long-lived radioactive isotopes that would present backgrounds at low-energy within the fiducial volume.

As we discuss in Sect. 3.1, the dominant background comes from ^{214}Pb, a daughter of omnipresent ^{222}Rn that originates from detector materials but is able to dissolve within the liquid xenon volume. Through the background reduction techniques described here, the low-energy background in XENON1T is <100 events/(t·year·keV), an unprecedentedly low background rate.

1.2 Physics Reach of XENON1T

As summarized in the previous section, combining its low energy threshold, large target mass, and low background rate—together with its particle identification and position reconstruction—XENON1T is an ultra-sensitive particle detector able to search for many rare-event processes. Using XENON1T we have set the most stringent WIMP dark matter constraints to date over a large mass range of 3–200 GeV [31, 33]. Recently we also reported the first observation of $2\nu ECEC$ (double electron capture) of ^{124}Xe; with a half-life $T_{1/2} \simeq 2 \times 10^{22}$ year, this is the rarest decay ever measured.

In this work, we use XENON1T to search for new physics using electronic recoil data. The signals we consider are (1) solar axions, (2) an enhanced neutrino magnetic moment, and (3) bosonic dark matter. In the following sections we describe these signals in more detail.

1.2.1 Solar Axions

Axions and the Strong CP Problem The SM describes with amazing precision the interactions of the fundamental particles of nature. One troubling aspect, however, is that in quantum chromodynamics (QCD) CP symmetry seems to be conserved experimentally even though it is not required theoretically. The CP-violating term in the QCD Lagrangian is [119]

$$\mathcal{L}_{QCD} \supset \bar{\theta}\frac{g_s^2}{32\pi^2}G\widetilde{G}, \tag{1.4}$$

where g_s is the strong coupling, and G ($\widetilde{G}$) is the gluon field (its dual). This term leads to a neutron electronic dipole moment (EDM) of

$$d_n \approx 5 \times 10^{-16}\,\bar{\theta}\text{ e cm}.$$

Experimentally, the neutron EDM is constrained to [133]

$$d_n < 0.18 \times 10^{-25}\text{ e cm},$$

implying a constraint on $\bar{\theta} \lesssim 10^{-10}$. The fact this parameter seems to "randomly" take on a value of $\sim$0 is referred to as the strong-CP problem.

As a solution to the strong CP problem, Peccei and Quinn postulated an additional global, chiral $U(1)$ symmetry (typically denoted $U(1)_{PQ}$) that when spontaneously broken gives rise to a pseudoscalar Nambu-Goldstone boson, the axion [134, 158, 159]. With the corresponding PQ current conserved apart from color and electromagnetic anomalies, the generalized Lagrangian added to the SM is given by [81]

$$\mathcal{L}_a \supset \frac{1}{2}(\partial_\mu a)^2 + \frac{a}{f_a}\frac{{g_s}^2}{32\pi^2}G\widetilde{G} + \frac{a}{f_a}\frac{e^2}{32\pi^2}\frac{E}{N}F\widetilde{F} + \frac{\partial_\mu a}{2f_a}\bar{f}c_f^0\gamma^\mu\gamma_5 f, \tag{1.5}$$

where a is the axion field, f_a is the symmetry-breaking energy scale of the PQ symmetry (also referred to as the axion decay constant), e is the charge of the electron, E is the electromagnetic anomaly factor, N is the color anomaly factor, f represents a fermion field (only one shown for brevity), and c_f^0 is a model-dependent axion-fermion coupling constant. The $aG\widetilde{G}$ term is what resolves the strong CP problem; after combing Eqs. (1.5) and (1.4), the axion field takes on a vacuum expectation value (VEV) that cancels out the $\bar{\theta}G\widetilde{G}$ term, replacing it with a dynamic variable (the axion field itself). The axion acquires a mass m_a given by [81]

$$m_a = \left(\frac{m_u m_d}{(m_u + m_d)^2}\frac{m_\pi^2 f_\pi^2}{f_a^2}\right)^{1/2} \simeq \frac{6 \times 10^6\text{ GeV}}{f_a}\text{ eV}, \tag{1.6}$$

where $m_{u/d}$ is the mass of the up/down quark, $m_\pi = 135$ MeV is the mass of the pion, and $f_\pi = 93$ MeV is the pion decay constant. This mass-decay constant relationship is valid for any axion model that solves the strong CP problem; such models are referred to as *QCD axions*.

Different QCD axion models predict different values of E, N, and c_f^0, leading to different couplings between axions and matter. In general, the axion-photon coupling $g_{a\gamma}$ is given by [67, 81, 115, 150]

$$\begin{aligned} g_{a\gamma} &= \frac{\alpha}{2\pi f_a}\left(\frac{E}{N} - \frac{2}{3}\frac{4m_d + m_u}{m_u + m_d}\right) \\ &\simeq \frac{\alpha}{2\pi f_a}\left(\frac{E}{N} - 1.92\right), \end{aligned} \tag{1.7}$$

where $\alpha = 1/137$ is the fine structure constant. The axion-electron coupling g_{ae} is given by [81]

$$g_{ae} = c_e^0 \frac{m_e}{f_a} + \frac{3\alpha^2 N m_e}{4\pi^2 f_a}\left(\frac{E}{N}\ln\frac{f_a}{m_e} - \frac{2}{3}\frac{4m_d + m_u}{m_u + m_d}\ln\frac{\Lambda}{m_e}\right), \tag{1.8}$$

where the first term is the model-dependent tree-level coupling and the second is radiatively induced, with m_e the mass of the electron and Λ the QCD cutoff scale. The axion-nucleon couplings g_{ap} and g_{an} are given by [79, 115, 150]

$$g_{ap} \simeq \frac{m_p}{f_a}(-0.47 + 0.88c_u^0 - 0.39c_d^0) \tag{1.9}$$

$$g_{an} \simeq \frac{m_n}{f_a}(-0.02 + 0.88c_d^0 - 0.39c_u^0), \tag{1.10}$$

where $m_p \simeq m_n$ are the masses of the proton/neutron and the higher-order terms from the other quark generations are ignored. Note that all axion-matter couplings contain a $1/f_a$ term and thus relate directly to the axion mass. Lighter axions are more weakly coupled and have longer decay constants.

The original axion model [158, 159], which predicted $f_a \sim$ the electroweak scale, has long been ruled out; however, "invisible" or ultra-light axions still viable today decouple the PQ symmetry breaking scale from the electroweak scale. There are two general benchmark models [81]:

- The Dine-Fischler-Srednicki-Zhitnitsky (DFSZ) type, where the SM fermions are charged under the new $U(1)_{PQ}$ symmetry, and thus there are tree-level couplings between axions and fermions [87, 165].
- The Kim-Shifman-Vainshtein-Zakharov (KSVZ) type, where there are no tree-level couplings to leptons and instead a new heavy quark charged under the PQ symmetry [118, 149].

We first consider the DFSZ-type axion, which includes two Higgs doublets charged under the PQ symmetry. These Higgs lead to an additional parameter in the DFSZ-type model beyond the axion mass, β_{DFSZ}, given by

$$\tan(\beta_{\text{DFSZ}}) = \left(\frac{X_u}{X_d}\right)^{1/2}, \tag{1.11}$$

where $X_{u,d}$ are the PQ charges of the up/down quarks. Note that conventions vary in the literature, with $X_u \leftrightarrow X_d$ sometimes swapped in Eq. (1.11) [81, 152]. These Higgs can couple to SM fermions in two different ways, leading to two sub-models of DFSZ type axions, DFSZ-I and DFSZ-II. We just quote the results here, referring to [81, 150] for details. Recalling the model dependence of Eq. (1.5), the DFSZ-I model predicts $E = 8$, $N = 3$, $c_{u_i} = \frac{1}{3}\sin^2\beta_{\text{DFSZ}}$, $c_{d_i} = c_{e_i} = \frac{1}{3}\cos^2\beta_{\text{DFSZ}}$, where the index i represents the three generations of the respective fermion. In the DFSZ-II model, the electromagnetic anomaly and axion-electron constants are replaced with $E = 2$ and $c_{e_i} = -\frac{1}{3}\sin^2\beta_{\text{DFSZ}}$, respectively. These values can be plugged into Eqs. (1.7)–(1.10) to give the respective couplings between DFSZ-type axions. It is also noteworthy that in any grand unified theory (GUT) combining the electroweak and strong interactions at high energies, $E/N = 8/3$ are the only allowed anomaly factors; thus, DFSZ-I is consistent with a GUT [81, 119].

In a KSVZ-type model, also called a *hadronic* axion model, none of the SM fermions are charged under the PQ symmetry. The axion-photon coupling remains due to the chiral anomaly, and thus loop diagrams allow for a suppressed but nonzero axion-electron coupling. The couplings to the SM quarks vanish, but a coupling to nucleons is still present due to mixing of the axion with the pion [119]. In the original, simplest KSVZ model $E = 0$, but in general E/N can take on many values [80]. To summarize, in KSVZ-type models, $c_{u_i} = c_{d_i} = c_{e_i} = 0$, and E/N is essentially a free parameter, albeit with some constraints not particularly relevant in this context. We will take a benchmark value of $E = 0$ for the majority of KSVZ-related plots in this work.

Since the axion is predicted to be massive, stable, and weakly interacting, it is also a very well-motivated dark matter candidate. Axions produced in the early universe through both thermal and non-thermal processes, such as the misalignment mechanism, could contribute to the dark matter density we observe (gravitationally) today [119]. Axionic dark matter would be *cold*, i.e. non-relativistic, similar to the WIMP model and consistent with precision cosmology measurements [14, 119]. In order to be dark matter, cosmological and astrophysical bounds constrain the axion the mass to be small (typically $\ll$ keV) [3, 65, 86, 138, 140]. With axions possibly solving two of the major questions in particle physics today, there is a large effort to search for these light, elusive, and intriguing particles.

Solar Axions in XENON1T On account of the astrophysical/cosmological mass constraint of $\ll$ keV, *dark matter* axions produced in the early Universe cannot be observed in XENON1T. However, due to the couplings to matter described in the previous section, axions can be produced in other processes, like in our own Sun.

These *solar* axions would emerge with—and in turn deposit—energies in the keV range [129, 141, 153], the precise energies to which XENON1T was designed to be most sensitive. An observation of solar axions would be groundbreaking evidence of beyond-the-SM physics, but would not by itself be sufficient to draw conclusions about axionic dark matter.

With couplings to photons, electrons, and nucleons, there are multiple stellar interactions that can produce axions. We consider three such production mechanisms contributing to the total solar axion flux: (1) Atomic recombination and deexcitation, Bremsstrahlung, and Compton (ABC) interactions [84, 141], (2) a mono-energetic 14.4 keV M1 nuclear transition of ^{57}Fe [129], and (3) the Primakoff conversion of photons to axions in the Sun [82, 139]. The ABC flux scales with the axion-electron coupling g_{ae} as

$$\Phi_a^{\mathrm{ABC}} \propto g_{ae}^2 \tag{1.12}$$

and was taken from [141]. The ^{57}Fe flux scales with an effective axion-nucleon coupling $g_{\mathrm{an}}^{\mathrm{eff}} \equiv -1.19 g_{\mathrm{an}}^0 + g_{\mathrm{an}}^3$ and is given by Alessandria et al. [16] and Andriamonje et al. [20]

$$\Phi_a^{^{57}\mathrm{Fe}} = \left(\frac{k_a}{k_\gamma}\right)^3 \times 4.56 \times 10^{23} (g_{\mathrm{an}}^{\mathrm{eff}})^2 \ \mathrm{cm}^{-2}\,\mathrm{s}^{-1}, \tag{1.13}$$

where

$$g_{an}^0 = \frac{1}{2}(g_{ap} + g_{an}) \tag{1.14}$$

$$g_{an}^3 = \frac{1}{2}(g_{ap} - g_{an}) \tag{1.15}$$

and k_a/k_γ is momentum ratio of the axion and photon, and is unity in the limit of a massless axion. The Primakoff flux scales with the axion-photon coupling $g_{\mathrm{a}\gamma}$ and is given by [122]

$$\begin{aligned}\frac{d\Phi_a^{\mathrm{Prim}}}{dE_a} &= \left(\frac{g_{\mathrm{a}\gamma}}{\mathrm{GeV}^{-1}}\right)^2 \left(\frac{E_a}{\mathrm{keV}}\right)^{2.481} e^{-E_a/(1.205\ \mathrm{keV})} \\ &\quad \times 6 \times 10^{30}\ \mathrm{cm}^{-2}\,\mathrm{s}^{-1}\,\mathrm{keV}^{-1},\end{aligned} \tag{1.16}$$

where E_a is the energy of the axion.

The expected flux from solar axions with $g_{ae} = 3 \times 10^{-12}$, $g_{\mathrm{a}\gamma} = 2 \times 10^{-10}$ GeV^{-1}, and $g_{\mathrm{an}}^{\mathrm{eff}} = 10^{-6}$ is shown in Fig. 1.3. These coupling values are arbitrarily chosen so that the fluxes be of the same order, and do not correspond to any particular QCD axion model. The chosen couplings are also roughly consistent with the current (direct) upper limits, highlighting that if axions exist they could be copiously produced in the Sun. While there could be a huge solar axion flux passing

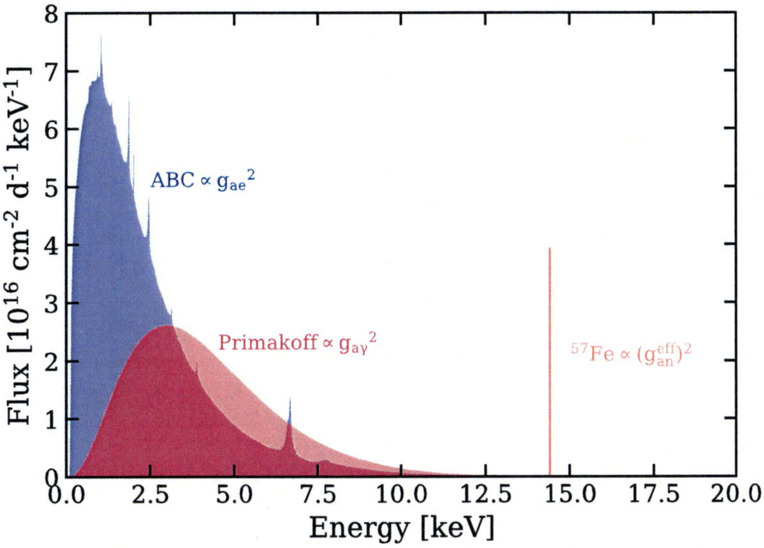

Fig. 1.3 Summary of the three-component solar axion flux. The arbitrary couplings chosen are $g_{ae} = 3 \times 10^{-12}$, $g_{a\gamma} = 2 \times 10^{-10}$ GeV^{-1}, and $g_{an}^{\text{eff}} = 10^{-6}$

through the Earth at all times, one needs a scattering cross section to determine how often an axion would interact.

All three flux components could be detected in XENON1T via the axioelectric effect—the axion analog to the photoelectric effect—which has a cross section that scales with the axion-electron coupling g_{ae} and given by Dimopoulos et al. [83, 85] Pospelov et al. [137], and Alessandria et al. [16]

$$\sigma_{ae} = \sigma_{pe} \frac{g_{ae}^2}{\beta} \frac{3E_a^2}{16\pi\alpha m_e^2} \left(1 - \frac{\beta^{2/3}}{3}\right), \tag{1.17}$$

where σ_{pe} is the photoelectric cross section in xenon and β and E_a are the velocity and energy of the axion, respectively. This cross section is shown in Fig. 1.4. For $m_a \lesssim 100$ eV, the axioelectric cross section is independent of axion mass; thus, the interaction rate for axions below this threshold depends only on the axion-matter couplings and not directly on the mass (though the couplings themselves are related to m_a in a model-dependent way, as discussed above). For the rest of this work we will assume $m_a < 100\,\text{eV}$ for solar axions.

Since solar axions would interact with the atomic electrons, their signature is an electronic recoil. The total event rate in a xenon detector is given by the convolution of the flux and cross section, and is shown in Fig. 1.5. We are thus able to constrain the values of $|g_{ae}|$ (ABC), $\left|g_{ae}g_{an}^{\text{eff}}\right|$ (^{57}Fe), and $|g_{ae}g_{a\gamma}|$ (Primakoff)[2] using

[2] We drop the absolute value notation for the remainder of this work.

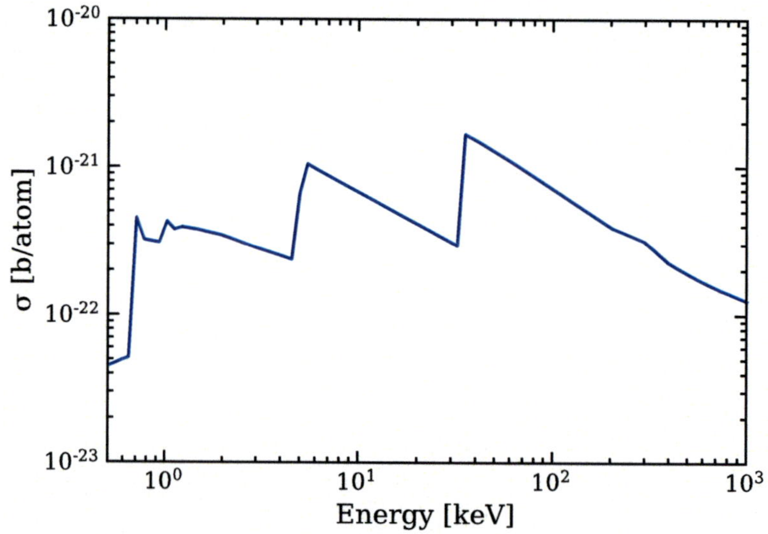

Fig. 1.4 Axioelectric cross section for $m_a \lesssim 100$ eV and $g_{ae} = 3 \times 10^{-12}$

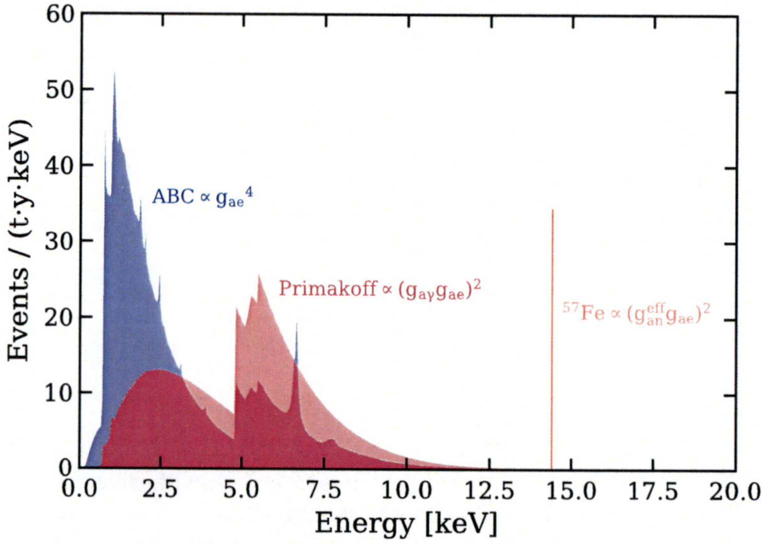

Fig. 1.5 Summary of the three-component solar axion rate in a LXe detector. The arbitrary couplings chosen are $g_{ae} = 3 \times 10^{-12}$, $g_{a\gamma} = 2 \times 10^{-10}$ GeV^{-1}, and $g_{an}^{\text{eff}} = 10^{-6}$

XENON1T. Lest we implicitly assume any particular axion model, we consider these three observables independently in the analysis. That is, since in principle it is possible for all three components to be present at the same time, our solar axion model (described in more detail in Chap. 3) includes three unconstrained parameters for the different components: ABC, Primakoff, and ^{57}Fe. The results of the three-component analysis will give constraints on the axion couplings, which could then be used to constrain the axion mass under different QCD model assumptions. This model-independent approach also implies that the results hold generally for solar axion-like particles, which do not have strict relationships between the couplings, as described in Sect. 1.2.3.

As an aside we note that after the initial release of this work it was shown that the we should also include the *inverse Primakoff effect*, in which an incoming axion scatters off the atomic electric field and converts to a $\sim$ keV photon, which would subsequently be photoabsorbed and produce an ER of the same energy [78, 100]. This cross section scales with $g_{a\gamma}$ and increases the expected axion interaction in XENON1T. We ignore this channel for this work, but note that it is an important effect to account for in future related analyses.

1.2.2 Neutrino Magnetic Moment

In the SM, neutrinos are massless and therefore lack a magnetic dipole moment. However, the observation of neutrino oscillation tells us that neutrinos have mass and the SM must be (at least) minimally extended, implying a magnetic moment of $\mu_\nu \sim 10^{-20}\,\mu_B$ [58, 96], where μ_B is the Bohr magneton. Larger values of μ_ν have been considered theoretically and experimentally [7, 57, 58]. Interestingly, in addition to providing evidence of beyond-SM physics, the observation of a $\mu_\nu \gtrsim 10^{-15}\,\mu_B$ would suggest that neutrinos are Majorana fermions [58]. Currently the most stringent direct detection limit is $\mu_\nu < 2.8 \times 10^{-11}\,\mu_B$ from Borexino [7], and indirect constraints based on the cooling of globular cluster and white dwarfs are an order of magnitude stronger at $\sim 10^{-12}\,\mu_B$ [44, 90, 152].

An enhanced magnetic moment would increase the neutrino scattering cross-sections at low energies (on both electrons and nuclei), and thus could be observable by low-threshold detectors such as XENON1T. Here we only consider the enhancement to elastic scattering on electrons (producing ER signals), given by Vogel and Engel [156]

$$\frac{d\sigma_\mu}{dE_\text{r}} = \mu_\nu^2 \alpha \left(\frac{1}{E_\text{r}} - \frac{1}{E_\nu} \right), \tag{1.18}$$

where E_r is the electronic recoil energy and E_ν is the energy of the neutrino. Note that Eq. (1.18) assumes free electrons; small corrections need to be made for the electron binding energies at $O(\text{keV})$ energies.

We search for an anomalous magnetic moment using solar neutrinos, whereby this signal would enhance the SM solar neutrino scattering rate at low energies. In calculating the expected signal rate we must first discuss the rate from SM processes. The differential solar neutrino flux is shown in Fig. 1.6, where the spectra were taken from [50] and the normalizations from [112]. This flux must then be convolved with

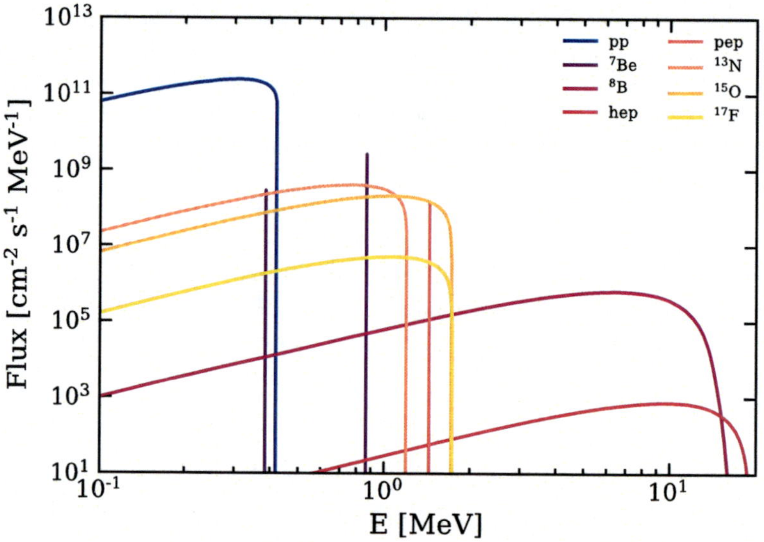

Fig. 1.6 Solar neutrino flux at the Earth, dominated by the low-energy *pp* reaction

the differential cross section for neutrino-electron elastic scattering, accounting for neutrino oscillations. The relevant cross sections (assuming free electrons) are [152]

$$\frac{d\sigma_{\text{SM}}(\nu_e e)}{dE_r} = \frac{G_F^2 m_e}{2\pi E_\nu}\left[4s_w^4(2E_\nu^2 + E_r^2 - E_r(2E_\nu + m_e)) - 2s_w^2(E_r m_e - 2E_\nu^2) + E_\nu^2\right],$$
$$\frac{d\sigma_{\text{SM}}(\nu_{\mu,\tau} e)}{dE_r} = \frac{G_F^2 m_e}{2\pi E_\nu}\left[4s_w^4(2E_\nu^2 + E_r^2 - E_r(2E_\nu + m_e)) + 2s_w^2(E_r m_e - 2E_\nu^2) + E_\nu^2\right], \tag{1.19}$$

where E_r is the recoil energy, E_ν is the neutrino energy, s_w is the sine of the Weinberg angle, and G_F is the Fermi constant. Since we are interested in keV-scale energy transfers, however, the electron wave functions are expected play a role. We approximate this effect by applying a step-function according to the accessible electrons for a given energy transfer, where the steps occur at the electron binding energies.

In order for a neutrino to transfer an energy $E_r \sim 1\,\text{keV}$ to an electron, it must have a minimum energy of

$$E_\nu^{\text{min}} = \frac{1}{2}\left(E_r + (E_r^2 + 2E_r m_e)^{1/2}\right) \sim 16\,\text{keV},$$

implying that near the XENON1T threshold, the *pp* neutrinos dominate. Assuming the Large Mixing Angle Mikheyev-Smirnov-Wolfenstein (LMA-MSW) model for neutrino oscillations, the ν_e survival probability at the Earth for these low-energy

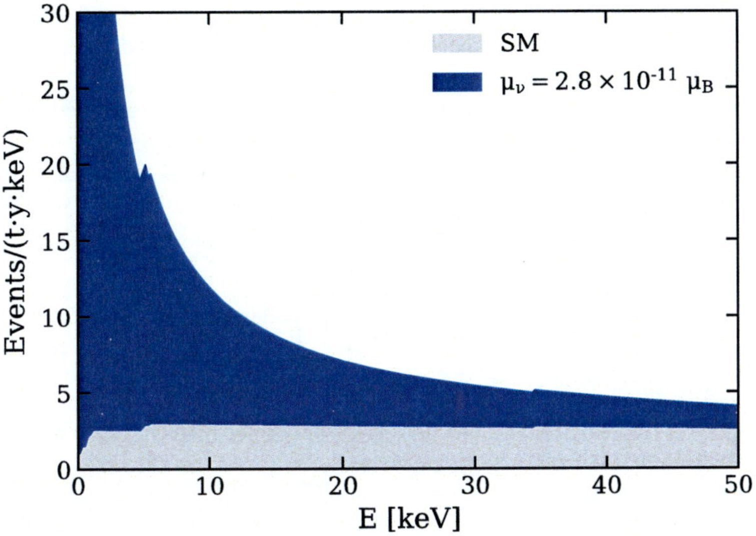

Fig. 1.7 Solar neutrino rate in liquid xenon detectors assuming only SM interactions (gray) and in the case of a large neutrino magnetic moment μ_ν (navy). In both cases, this spectra are dominated by the pp neutrinos

neutrinos is dominated by matter effects in the Sun and given by $p \sim 0.57$. Combining the flux in Fig. 1.6, cross sections in Eq. (1.19), the step function approximation, and this survival probability, the expected neutrino-electron scattering rate in XENON1T is shown in gray in Fig. 1.7. With a total rate of ~50 events/t/year below 20 keV, the SM background is quite subdominant in XENON1T, as discussed in more detail in Chap. 3.

To calculate the expected enhancement to the SM solar neutrino rate from an anomalous magnetic moment, we repeat the same procedure as in the SM case but replacing the cross section with Eq. (1.18). This enhancement is also shown in Fig. 1.7 for $\mu_\nu = 2.8 \times 10^{-11}\ \mu_B$, the current upper limit from Borexino, where one can see the steep increase in the spectrum of the spectrum at low energies. While the SM solar neutrino rate is too low to be measured with XENON1T (though we do treat it as a background as discussed in Chap. 3), an enhancement in the rate due to a large magnetic moment could be observable in XENON1T.

1.2.3 Bosonic Dark Matter

The third class of signals we consider is bosonic dark matter, specifically pseudoscalar bosons and vector bosons.

Axion-like particles (ALPs), like QCD axions, are pseudoscalar bosons, but with decay constant and particle mass (Eq. (1.6)) decoupled from each other and instead taken as two independent parameters. This decoupling allows for ALPs to take on higher masses than QCD axions; however, it also implies that ALPs do not solve the strong CP problem.

ALPs are viable dark matter candidates [45], and could be absorbed in XENON1T via the axioelectric effect (Eq. (1.17)) like their QCD counterparts. Assuming ALPs are non-relativistic and make up all of the local dark matter (density $\rho \sim 0.3\,\text{GeV/cm}^3$ [77]), the expected signal is a mono-energetic peak at the rest mass of the particle, m_a, with an event rate given by [46, 137]

$$R \simeq \frac{1.5 \times 10^{19}}{A} g_{ae}^2 \left(\frac{m_a}{\text{keV}/c^2}\right)\left(\frac{\sigma_{\text{pe}}}{\text{b}}\right) \text{kg}^{-1}\text{day}^{-1}, \tag{1.20}$$

where A is the average atomic mass of the detector medium ($A \approx 131$ u for xenon). The rate coefficient from our calculation is consistent with [63] for the dark matter density used in this work.

In addition to the pseudoscalar ALPs, XENON1T is also sensitive to vector bosonic dark matter, of which dark photons are a common example. Dark photons can couple weakly with SM photons through kinetic mixing [99] and be absorbed with cross section σ_{V} given by [19]

$$\sigma_{\text{V}} \simeq \frac{\sigma_{\text{pe}}}{\beta}\kappa^2, \tag{1.21}$$

where σ_{pe}, α, and β are the same as in Eq. (1.17), and κ parameterizes the strength of kinetic mixing between the photon and dark photon. Similarly to Eq. (1.20), by following the calculation in [137], the rate for non-relativistic dark photons in a detector reduces to

$$R \simeq \frac{4.7 \times 10^{23}}{A} \kappa^2 \left(\frac{\text{keV}/c^2}{m_{\text{V}}}\right)\left(\frac{\sigma_{\text{pe}}}{\text{b}}\right) \text{kg}^{-1}\,\text{day}^{-1}, \tag{1.22}$$

where m_{V} is the rest mass of the vector boson.

To summarize, both pseudoscalar and vector bosonic dark matter would leave the same signature in the detector, just with different interpretations due to the different cross sections. We thus combine these two signals as a single "peak search", where will look for excess events consistent with line sources of a single energy, and draw conclusions for both types of bosonic dark matter.

Chapter 2
Event Reconstruction and Selection

Abstract This chapter is devoted to the reconstruction and selection of events used for this analysis. Section 2.1 details how we reconstruct the deposited energy of electron recoil events in XENON1T and summarizes the energy resolution of the detector. Section 2.2 focuses on the energy threshold for this analysis, making use of a combination of data-driven and pure simulation methods to extract a total detection efficiency as a function of reconstructed energy. Section 2.3 describes the selection of events used for this analysis and the signal acceptance of the applied cuts. The main results of this chapter provide understanding of how the XENON1T detector reconstructs events on a fundamental level and are important inputs for the modeling and interpretation of the data that come in later chapters.

2.1 Energy Reconstruction and Resolution

2.1.1 Energy Reconstruction

As discussed in Chap. 1, both the S1 and S2 signals are proportional to the energy of an event. For nuclear recoils, energy loss due to heat goes undetected, leading to a "quenching factor" that needs to be accounted for in such events. Electronic recoils, on the other hand, lose a negligible amount of energy to heat, meaning that the total energy is ultimately distributed to the two channels that we can detect: scintillation (S1) and ionization (S2). This allows for a robust energy reconstruction that holds over a large energy range.

Since the energy of an ER can go into two channels, we can write the total energy deposited E as

$$E = W(n_{ph} + n_e), \tag{2.1}$$

where $n_{ph/e}$ are the numbers of photons/electrons released after recombination and $W = 13.7$ eV/quanta is the "W-value" denoting the average energy required to create one "quantum" (photon or electron) [76]. With $S1 \propto n_{ph}$ and $S2 \propto n_e$, we can reconstruct the total energy if we know the light and charge yields. As these yields

E. Shockley, *Study of Excess Electronic Recoil Events in XENON1T*,
Springer Theses, https://doi.org/10.1007/978-3-030-87752-1_2

are themselves energy-dependent [151], this method relies on precise calibrations at many energies and likely a deep theoretical understanding of the microphysics involved. While this is a difficult task, we can circumvent it in ER events due to the fact we detect both channels of energy loss, implying that the light and charge yields are linearly anti-correlated. Since the total energy is proportional to the sum of n_{ph} and n_e, if the yield (per unit energy) of one channel increases then that of the other must decrease in linear fashion.

The S1 and S2 signals depend on many other factors beyond the n_{ph} and n_e produced at the interaction site, however. An emitted photon must hit the photocathode of a PMT in order to be detected; the probability of this happening depends on the solid angle subtended by the PMTs and the number of reflections required to hit a PMT, leading to a spatial dependence from detector geometry. Further, since the recombination fraction depends on the local electric field $V(x, y, z)$, non-uniformities of the applied field V_0 can also lead to position-dependent n_{ph} for the same energy deposited. We denote the average photon yield as PY. A photon that does reach a photocathode produces a photoelectron with probability given by the quantum efficiency ϵ_{QE} of that PMT. There is additional efficiency loss due to the charge collection efficiency, ϵ_{CE}, as well as the PMT-specific DAQ thresholds to read out a signal, ϵ_{trig}. The effects from ϵ_{QE} and ϵ_{trig} can also have some spatial dependence, albeit less so than the detector geometry, due to PMT variability. Therefore, for an energy deposited E, the mean S1 signal is given by

$$S1(E, x, y, z, V_0) = PY(E, V(V_0, x, y, z)) \cdot \epsilon_{QE}(x, y, z) \cdot \epsilon_{CE}(x, y, z) \cdot \epsilon_{\text{trig}}(x, y, z) \tag{2.2}$$

While we cannot determine each term in Eq. (2.2) precisely, we can correct for the spatial dependence of S1 signals using calibration. In this vein a $^{83\text{m}}$Kr calibration source was injected into the gaseous xenon and dissolved uniformly in the liquid volume. This source exhibits a two-step decay of known energy (32.1 + 9.4 keV) with a half-life of ~150 ns, making it straightforward to discriminate calibration events from background events. Since the energy of each $^{83\text{m}}$Kr event is the same, we can calibrate the position-dependence of the mean S1 signal using this source. A correction factor is applied to all S1s to match the mean at the center of the detector, with values shown in Fig. 2.1. We refer to the corrected S1 as $cS1$; as this accounts for spatial dependence, this is the variable used in energy reconstruction.

The S2 signal must also be corrected for spatial effects. As discussed in Chap. 1, electronegative impurities in the liquid xenon can capture electrons during the drift, meaning that for a fixed n_e the observed S2 has a strong z dependence. Recalling that the electron lifetime τ quantifies this loss (see (1.3)), a corrected S2 signal is given by

$$cS2^0 = S2 \cdot e^{t/\tau}, \tag{2.3}$$

where t is the drift time (time between S1 and S2) and we use $cS2^0$ because there will be more corrections that follow. Since the electron lifetime is highly dependent

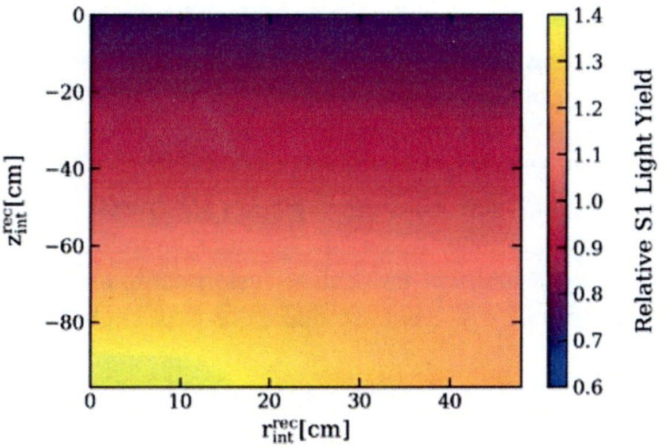

Fig. 2.1 Correction applied to S1 signals depending on their position in the detector, averaged over the ϕ dimension for simplicity

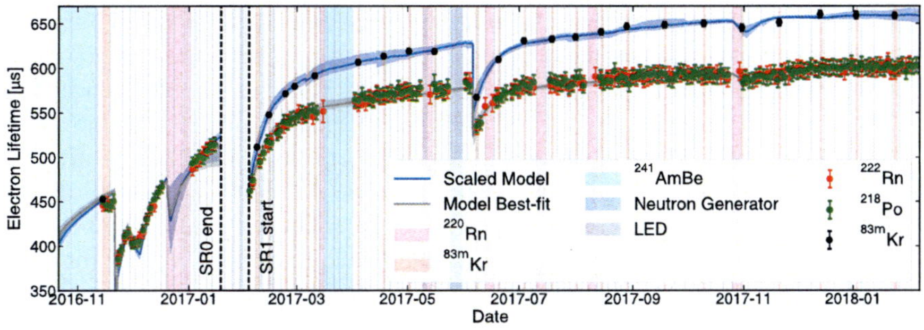

Fig. 2.2 Electron lifetime as a function of time in XENON1T, taken from [37]

on xenon purity, τ has a strong time dependence that needs to be accounted for. The time dependence of τ can be found in Fig. 2.2, where it was calibrated using ^{83m}Kr data along with other sources. The offset between the different sources is not understood at this time, but it is scales with xenon purity [37]. Since we are focused on low-energy signals, we use the electron lifetime calculated with ^{83m}Kr data. After the electron lifetime correction, the S2 signal is further corrected for x-y dependencies arising due to (1) sagging of the anode mesh in the center of the detector, leading to relatively smaller S2s there and (2) electric field and PMT effects, similar to the S1 correction. This correction is also calibrated using ^{83m}Kr data. Since the S2 signal is extremely localized in the top array, PMT effects can play a large role in the total S2 area; therefore, to reduce this source systematic uncertainty only the corrected S2 signal in the bottom PMT array is used for energy reconstruction. That is:

$$cS2_b = cS2_b^0 \cdot C_b(x, y) \tag{2.4}$$

where $cS2_b^0$ represents the corrected S2 size in the bottom array after electron lifetime correction, and $C_b(x, y)$ is the x-y S2 correction for the bottom array. We use $cS1$ and $cS2_b$ for all energy reconstruction in XENON1T.

Since $cS1$ and $cS2_b$ are proportional to the number of quanta produced, we introduce two proportionality constants $g_1 = \overline{cS1}/n_{ph}$ and $g_2 = \overline{cS2_b}/n_e$. These detector-specific "gain constants" give us the average *detected* signal ($cS1$ and $cS2_b$) for a given number of quanta (photons and electrons) *produced*. Once we know g_1 and g_2, which we can extract using calibration data, we can easily reconstruct the energy of any event. To make this explicit, we can rewrite Eq. (2.1) as

$$Q_y = -\frac{g_2}{g_1}L_y + \frac{g_2}{W}, \tag{2.5}$$

where $Q_y = \overline{cS2_b}/E$ is the charge yield and $L_y = \overline{cS1}/E$ is the light yield. Note that Eq. (2.5) is a linear equation with two unknowns. Thus, by measuring the yields for several calibration sources of known energies, we can then perform a linear fit in the space of Q_y-vs-L_y to extract the values of g_1 and g_2. This is shown below, where we consider separately two different science runs. The first, called Science Run 1 (SR1) is the primary dataset of XENON1T and includes the longest and most stable period of science data. The second, called Science Run 2 (SR2), is primarily an R&D dataset with changing detector conditions due to improvements to purification system; however, it also includes a stretch of science data that is used in this analysis. Perhaps more importantly, during SR2 an additional, low-energy calibration source in ^{37}Ar was deployed. We discuss this important calibration point in more detail in the SR2 section below.

Energy Calibration in SR1 As discussed in Sect. 2.3 below, for this analysis we consider an energy range of (1, 210) keV. Accordingly, we focus on low-energy calibration points when extracting g_1 and g_2 so that we can safely ignore saturation effects at high energies. In SR1 we use five calibration points:

1. ^{83m}Kr, an injected calibration source that dissolves uniformly in the liquid volume and decays via a two-step scheme, as described above. Since the two S2s from these decays are often unresolved, for this study consider ^{83m}Kr events where the 32- and 9-keV decays are merged into a single S1.
2. ^{131m}Xe, an activated isotope of xenon that decays via internal conversion and deposits a total energy of 163.9 keV. Being a xenon isotope, this source is uniformly distributed in the TPC. It is abundantly produced through neutron capture during neutron calibration campaigns, after which it decays with a half-life of ~10 days.
3. ^{129m}Xe, an activated isotope of xenon, similarly to ^{131m}Xe, that is produced during neutron calibrations. It deposits total energy of 236.1 keV.
4. ^{214}Bi, a daughter of ^{222}Rn, which emanates from detector materials and dissolves uniformly in the detector. ^{214}Bi can decay via emission of an β to an excited state of ^{214}Po, whose (resolved) γ deposits a total energy of 609.3 keV.

5. ^{60}Co, a background originating from detector materials such as the PMTs, which can β-decay to an excited state of ^{60}Ni, which in turn decays via as 1333-keV γ. The source is not distributed uniformly in the TPC.

Figure 2.4 shows the calibration data in S2-vs-S1 space, where the five calibration points can be seen above a diffuse background. The elliptical shapes of the mono-energetic sources in this space is indicative of the anti-correlated behavior of the light/charge yields. To extract the yields we maximize (for each calibration point) a binned log-likelihood for a rotated 2D Gaussian, given by

$$L(\mu_1, \mu_2, \sigma_1, \sigma_2, \theta) = \sum_{b_1, b_2} N_{\text{obs}}(b_1, b_2)\Bigg(- \frac{((b_1 - \mu_1)\cos\theta - (b_2 - \mu_2)\sin\theta)^2}{2\sigma_1^2} - \frac{((b_1 - \mu_1)\sin\theta + (b_2 - \mu_2)\cos\theta)^2}{2\sigma_2^2} - \ln(\sigma_1 \sigma_2)\Bigg), \tag{2.6}$$

where μ_i and σ_i are the means and standard deviations of the cS_i distribution with $i \in \{1, 2\}$, θ is the rotation angle to account for the correlation between light and charge yields, $b_{1,2}$ are the bin-centers of the 2D histogram shown in Fig. 2.3, and $N_{\text{obs}}(b_1, b_2)$ is the corresponding bin content for each bin. We choose small enough

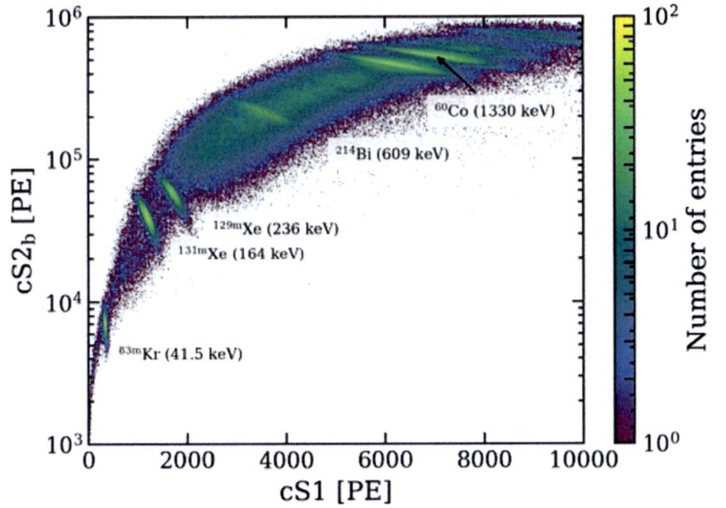

Fig. 2.3 An S2-S1 plot from SR1 in XENON1T, where mono-energetic sources are visible as the elliptical peaks. The labeled peaks are the ones used to calibrate the energy reconstruction in this analysis

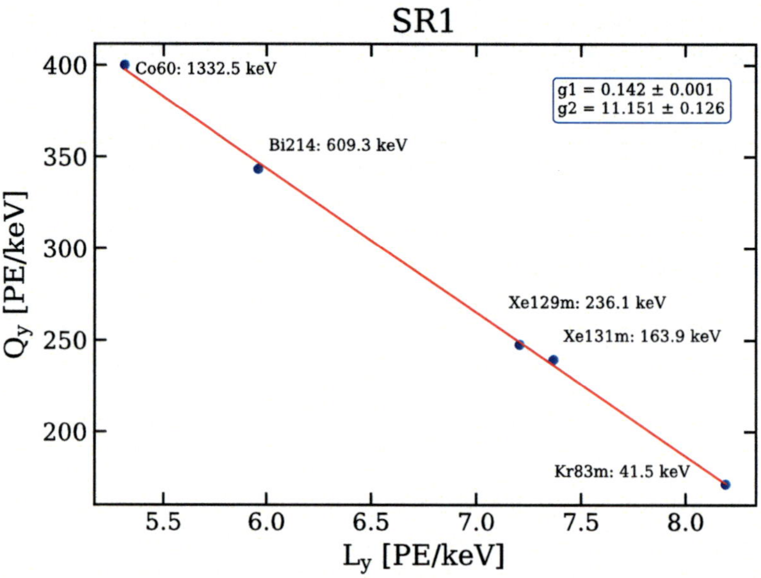

Fig. 2.4 Charge yield vs. light yield for the five monoenergetic ER sources considered in SR1. Statistical uncertainties are smaller than the data points. As expected, the yields correlate linearly, and g_1 and g_2 can be extracted using a simple fit

bin sizes so that we can ignore changes in the probability density function within each bin, and we ignore a term for the background since it is sub-dominant. After maximizing this likelihood, the best-fit $\mu_{1,2}$ along with the known corresponding energies give us the light/charge yields for each source, with statistical uncertainties propagated from the fit.

In Fig. 2.4, we show the resulting light/charge yields for the five sources used in SR1, along with a linear fit using Eq. (2.5). From this fit, we conclude that for SR1

$$g_1 = 0.142 \pm 0.001 \text{ PE/photon}$$
$$g_2 = 11.15 \pm 0.13 \text{ PE/electron.}$$

The uncertainty is that given by the statistical uncertainty propagated from the various fits. These g_1 and g_2 can then be used to reconstruct the energy of any event. As a self-consistency check, we show the spectrum of the SR1 calibration data used for this study in Fig. 2.5, where the three low-energy peaks are reconstructed as expected. We focus on the low-energy region because that is the most relevant for this analysis.

As an aside, it is noteworthy that the EXO-200 experiment has recently published a different value for W of 11.5 ± 0.5 eV [21]. While changing the W value would slightly modify our understanding of the light and charge yields, we note that this change would have no effect on the energy reconstruction discussed here. In fact,

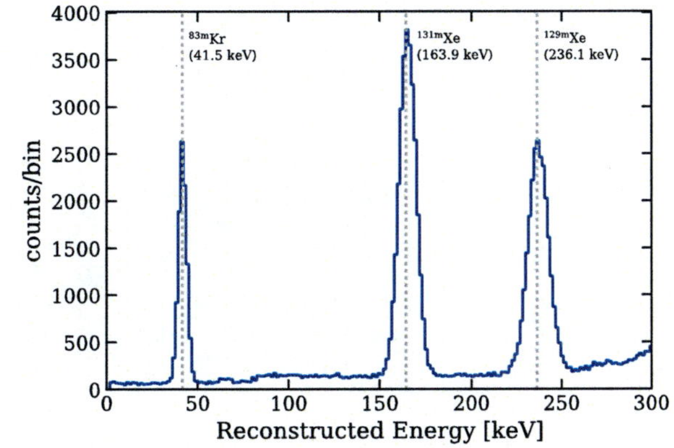

Fig. 2.5 Self-consistency check of the SR1 energy calibration, focusing on the low-energy region. The ^{83m}Kr, ^{131m}Xe, and ^{129m}Xe monoenergetic peaks are all reconstructed where expected

as long as W is constant with energy, any change its value would just be absorbed into the fitted values of g_1 and g_2. For example, when fitting the data in Fig. 2.4 with $W = 11.5$ eV, we have $g_1 = 0.119 \pm 0.001$ PE/photon and $g_2 = 9.36 \pm 0.106$ PE/electron, but the linear model describes the data just as well. It is clear by this that, while the values of the g_1 and g_2 constants change for different W values, the energy reconstruction itself would not be affected.

Energy Calibration in SR2 The SR1 data shown above is used for the main result of this analysis, but we also consider a dataset called SR2. During SR2 we performed an additional calibration campaign using ^{37}Ar. This gaseous source was injected into the xenon purification system, where it then dissolved uniformly within the liquid xenon volume. ^{37}Ar decays via electron capture to the ground state of ^{37}Cl, emitting X-rays in the process to fill the vacancy. Depending on vacant shell, this results in 2.82 keV (K-shell), 0.27 keV (L-shell), or 0.0175 keV (M-shell) X-rays, with decreasing branching ratios [55]. Focusing on the K-shell, this calibration allows us to check that the energy scale holds at 2.8 keV and validates the use of the combined energy scale down to our energy threshold of 1 keV.

Since ^{37}Ar decays deposit energies near our energy threshold—which is dominated by a requirement that at least 3 PMTs detect a signal within a 50 ns time window—the detection efficiency needs to be accounted for when extracting the light yield. We discuss the detection efficiency in detail later in this chapter, but for now we take the resulting S1 efficiency (Fig. 2.20, right) and modify our likelihood accordingly by adding an additional term $\ln(\epsilon_{\text{det}}(b_1))$ to Eq. (2.6), where $\epsilon_{\text{det}}(b_1)$ is the detection efficiency at the corresponding S1 bin center. The ^{37}Ar data and the result of this maximum likelihood fit is shown in Fig. 2.6, where the asymmetric behavior in S1 is due to this detection efficiency.

Due to different detector conditions in SR2 compared to SR1, we don't expect g_1 and g_2 to be the same. Thus we extract g_1 and g_2 for SR2 following the same approach but with ^{37}Ar included as described above. The resulting light/charge yield

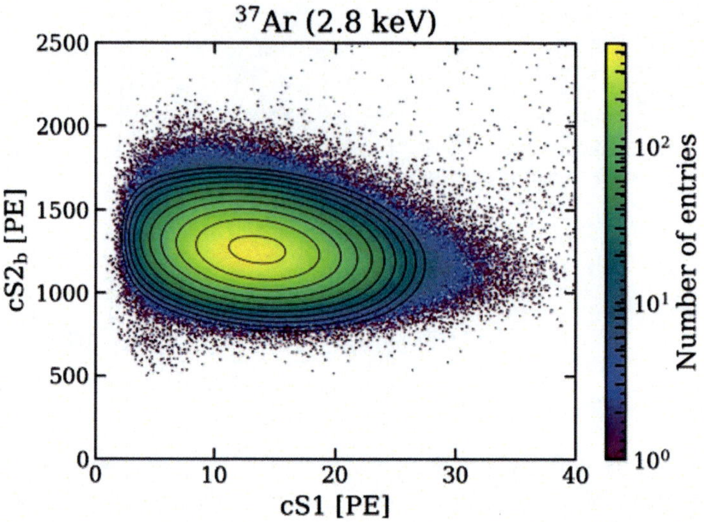

Fig. 2.6 ^{37}Ar calibration data in S1-S2 space, along with the contours from the maximum-likelihood fit (including detection efficiency) used to extract the light and charge yields

plot is shown in Fig. 2.7, from which we can conclude that our energy reconstruction holds down to 2.8 keV. This fit gives SR2 values of

$$g_1 = 0.1439 \pm 0.0004 \text{ PE/photon}$$
$$g_2 = 11.23 \pm 0.05 \text{ PE/electron.}$$

These g_1 and g_2 values are quite similar to the SR1 values.

It is important to stress again that we did not perform an ^{37}Ar calibration campaign during SR1, which is the main dataset we consider for this analysis. Thus, for this time region, the extrapolation of our energy reconstruction to low energies does not have a calibration point there. While it may seem concerning, this extrapolation is validated using SR2 data, where the energy reconstruction does hold to low energies. To check the systematic of including ^{37}Ar in SR2 but not SR1, we can also extract g_1 and g_2 values for SR2 ignoring the ^{37}Ar point. This gives

$$g_1{}^{sys} = 0.1438 \pm 0.0004 \text{ PE/photon}$$
$$g_2{}^{sys} = 11.24 \pm 0.07 \text{ PE/electron,}$$

which is very much consistent with the values *including* ^{37}Ar above. We thus conclude that the extrapolation of the energy reconstruction down to low energies in SR1 is robust, as expected.

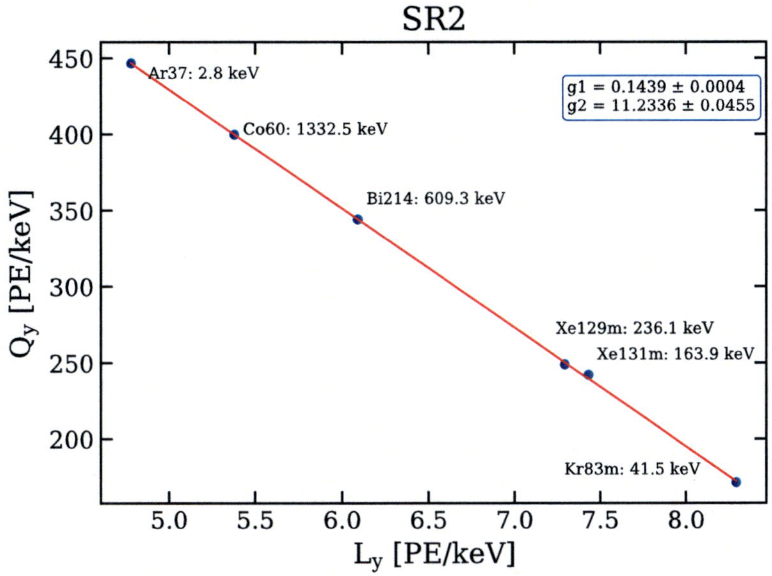

Fig. 2.7 Charge yield vs. light yield for the six monoenergetic ER sources considered in SR2. The yields (anti-) correlate linearly, and g_1 and g_2 can be extracted using a simple fit. The inclusion of ^{37}Ar is an important check that our energy reconstruction holds to low energies

2.1.2 Energy Resolution

We use an empirically-motivated function to model the energy resolution in this analysis. The monoenergetic sources are fitted as Gaussians and the resulting standard deviations saved. The standard deviation, σ as a function of energy is then fitted with

$$\sigma(E) = a \cdot \sqrt{E} + b \cdot E, \tag{2.7}$$

where E is the energy and a and b are constants. The fit yields $a = (0.310 \pm 0.004)\sqrt{\text{keV}}$ and $b = 0.0037 \pm 0.0003$, and is shown Fig. 2.8. We only consider the peaks <300 keV, since this is the range relevant for the analysis described in the following chapters.

This energy resolution is used to smear the theoretical signal/background models to account for detector effects; i.e., the resolution function converts spectra from deposited energy to reconstructed energy. This smearing is performed via

$$R_r(E_r) = \int_{E_t} R_t(E_t)\ f_G(E_r|\mu = E_t, \sigma = \sigma(E_t))\, dE_t, \tag{2.8}$$

where an r subscript denotes 'reconstructed', a t subscript denotes 'true', R_t is the true spectrum at a given (true) energy, R_r is the reconstructed spectrum at a

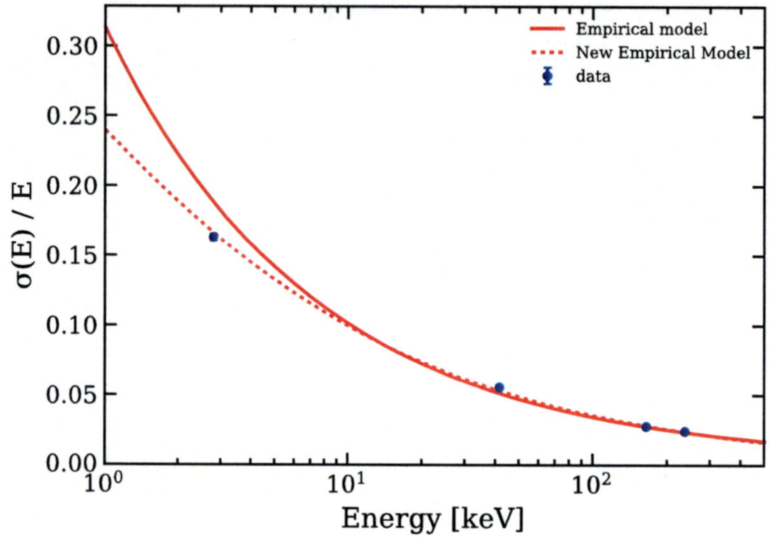

Fig. 2.8 Energy resolution in XENON1T

given (reconstructed) energy, f_G is the Gaussian probability density function (pdf), and $\sigma(E_t)$ is our energy resolution function from (2.7). For simplicity we assume Gaussian smearing for the entire region.

We note that the fit in Fig. 2.8 (red) mismodels the low-energy region slightly, where the observed ^{37}Ar resolution is better than the model by a few percent. Based on this, we also considered using the XENON microphysics model outlined in [37]; however, this model only holds to ~10 keV and also did not accurately predict the resolution of the ^{37}Ar data. We therefore use the empirical model described above for this analysis.

To study the effect of this choice for the resolution, we consider here an additional empirical fit model, shown in dashed in Fig. 2.8. We take two benchmark signal models from Chap. 1, an ABC solar axion and a 2-keV ALP, and smear their true spectra with the two resolution functions using (2.8). The results are shown in Fig. 2.9. For the solar axion spectrum, the difference between the two resolution models is negligible; the small differences observed in Fig. 2.9 are suppressed even further after accounting for the detection efficiency (discussed later in this chapter). Other continuous energy spectra like the neutrino magnetic moment will similarly be unaffected. For the ALP model in Fig. 2.9, the different resolution functions yield a ~10% difference in peak heights. This difference, while not negligible, is not so large as to be a concern for the analysis. We thus use the empirical model from (2.7) for the main results, as this is the same resolution function used for other analyses with XENON1T [34, 39]. To summarize, our choice of resolution function is expected to slightly affect the results of the low-mass bosonic dark matter search. We will revisit this systematic in the context of these results in Chap. 4.

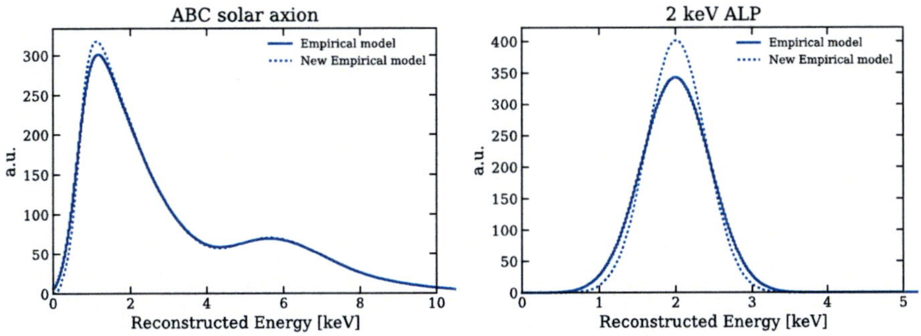

Fig. 2.9 Comparison of different resolution functions on an ABC solar axion model (left) and a 2 keV ALP model (right). The differences for the solar axion model are negligible after considering the detection efficiency. For the ALP model, the different resolution models would have a small but not negligible effect on the overall rate constraint

2.2 Detection Efficiency

For ER events, the focus of this analysis, the energy threshold is dominated by the *S1 detection efficiency*, the probability of detecting and correctly classifying an S1 signal. The analogous efficiency for S2s is ~100% for S2 $\gtrsim$ 200 PE; with ER events above 1 keV all having S2 > 500 PE, we can safely assume no efficiency loss due to S2 detection [37].

The energy threshold depends on the following factors:

- The photon detection efficiency, absorbed in the g_1 value extracted in Sect. 2.1.1.
- The response of the PMTs, i.e., the probability that a signal in a PMT rises above its detection threshold. We refer to this as the *single-photoelectron (SPE) acceptance*, which differs for each PMT and in principle is a function of time.
- The efficiency of the XENON1T reconstruction software at identifying S1 signals, based primarily on the arrival time of detected photons. We refer to this as the *reconstruction efficiency*.

As mentioned, the photon detection efficiency is included in the calibration of g_1, and so we focus on the SPE acceptance and S1 reconstruction efficiency contributions in this section.

For a given number of photons detected (i.e., that create photoelectrons) n_{phd}, the total detection efficiency is given by

$$\begin{aligned}\epsilon(n_{phd}) &= \mathrm{P}(S1|n_{phd}) \\ &= \sum_{n_h} \mathrm{P}(S1|n_h)\,\mathrm{P}(n_h|n_{phd}) \\ &= \sum_{n_h} \epsilon_{\mathrm{rec}}(n_h)\,\mathrm{P}(n_h|n_{phd}),\end{aligned} \tag{2.9}$$

where

- P($S1$) refers to the probability of observing an S1 signal in the reconstructed data.
- n_h is the number *hits*, photon signals that induce a voltage drop with amplitude exceeding a PMT's threshold. This threshold is PMT-dependent and is set at the level of the digitizer.
- $\epsilon_{\text{rec}}(n_h) \equiv \text{P}(S1|n_h)$ is the S1 *reconstruction efficiency*, the probability of observing an S1 given that n_h hits were detected. This term does not include any knowledge of the PMT response, but only of the reconstruction algorithms. $\text{P}(n_h|n_{phd})$ gives the probability of observing n_h hits for a given number of n_{phd} detected. This term includes the effect of the PMT response.

In the sections that follow, we calculate $\epsilon_{\text{rec}}(n_h)$ and $\text{P}(n_h|n_{phd})$. We employ two different techniques to estimate the S1 reconstruction efficiency in Sect. 2.2.2, while $\text{P}(n_h|n_{phd})$ can be determined from the SPE acceptance, described in the next section.

2.2.1 Single Photoelectron Acceptance

The trigger system of XENON1T has two levels in normal TPC mode. First, each PMT has its own *self-trigger* threshold, a channel-specific value, relative to baseline, that is set in the digitizer and determines when signals from that channel should be read out by the data acquisition (DAQ). PMT signals that do not reach the self-trigger threshold are not stored in the data, and therefore excluded from the event reconstruction. Such a threshold is needed so that the detector is not constantly triggered by noise, but as we detail below it does lead to some efficiency loss.

PMT signals that exceed their self-trigger thresholds are stored in memory in a Mongo database, in which the second level, the *software trigger*, determines if enough PMTs detected signals in a particular time region for it to be stored as an *event* [1, 35]. The first level is more important for studying the S1 detection efficiency, as the specific self-trigger threshold (along with PMT-specific properties like gain, geometry, and general response) for each PMT affects the SPE acceptance. The software trigger requires a comparatively large signal (usually an S2) above 200 PE, and so is not relevant here.

The XENON1T PMTs were calibrated using LEDs, which emitted visible photons that were guided inside the detector via optical fibers. Rather than relying on the self-trigger described above, for LED calibration the DAQ was externally triggered, in sync with the emission of the LED photons. In this way, with the expected photon arrival time given by the external trigger, the true distribution of single photoelectrons (SPEs) can be determined. The fraction of SPE signals that have amplitudes larger than the self-trigger threshold is the SPE acceptance. We thus are interested in the amplitude distribution of SPE signals.

The steps to calibrate the SPE acceptance are outlined in Fig. 2.10. The procedure is based off a model-independent approach to determine the SPE gain [145]. We use two different datasets to extract this value: (1) an externally triggered LED dataset, with total light emitted tuned to be small enough so that we are in the SPE regime, and (2) a "blank" dataset, which is triggered in the exact same way but with no light emitted into the detector. Due to the external trigger the time window of the LED signal is known, and so we can build the amplitude spectrum for that time region. For the LED dataset this spectrum is the convolution of LED-induced SPE, LED-induced two-or-more photoelectrons (2^+PE), noise, and dark counts (not LED-induced photoelectrons). The amplitude spectrum for the noise dataset includes contributions from noise and dark counts only. As an example, the spectra for both an LED and blank dataset for PMT channel 0 are shown in Fig. 2.10 in red and gray, respectively. The contribution 2^+PE is sub-dominant (by design), but it leads to the high-ADC count tail of the SPE distribution. We only consider positive amplitudes (relative to baseline), so there are no counts for amplitudes < 0.

Since the LED run includes SPE and 2^+PE contributions while the blank run does not, the normalization of the noise pedestal is different between the two. To correct for this, we make an assumption that no real SPE signals have amplitude less than a given threshold value t, and then scale the blank spectrum to have the same integral as the LED run below this threshold value. We vary t to study this systematic, but in Fig. 2.10 we use $t = 6$ ADC counts. To be explicit, we apply a correction c to the blank data, given by

$$c(t) = \frac{\sum_{i=0}^{t} A_{\mathrm{LED}}(i)}{\sum_{i=0}^{t} A_{\mathrm{blank}}(i)} \tag{2.10}$$

where i is the iterator over the amplitude bins, A_{LED} is the LED amplitude spectrum, and A_{blank} is the blank amplitude spectrum. The blank spectrum after correcting with c is shown in black in Fig. 2.10 (top), and is what will be used to extract the SPE spectrum.

The correction factor c is related to the *occupancy*, the average number of photoelectrons detected per trigger. Since we assume that no SPE signals have an amplitude $\leq t$, we are effectively measuring with this correction the probability of detecting 0 photoelectrons in a given trigger. This is a Poisson process [105], with probability mass function

$$P(k) = \frac{\lambda^k e^{-\lambda}}{k!} \tag{2.11}$$

where λ is the occupancy and k is the number of photoelectrons detected. With these assumptions, we observe that $c = P(k = 0)$, from which we have $\lambda = -\ln c$. Once we know λ, we can also estimate the contribution of 2^+PE (i.e., $P(k \geq 2)$) using Eq. (2.11):

$$P(k \geq 2) = 1 - P(k < 2) = 1 - e^{-\lambda}(1 + \lambda) \tag{2.12}$$

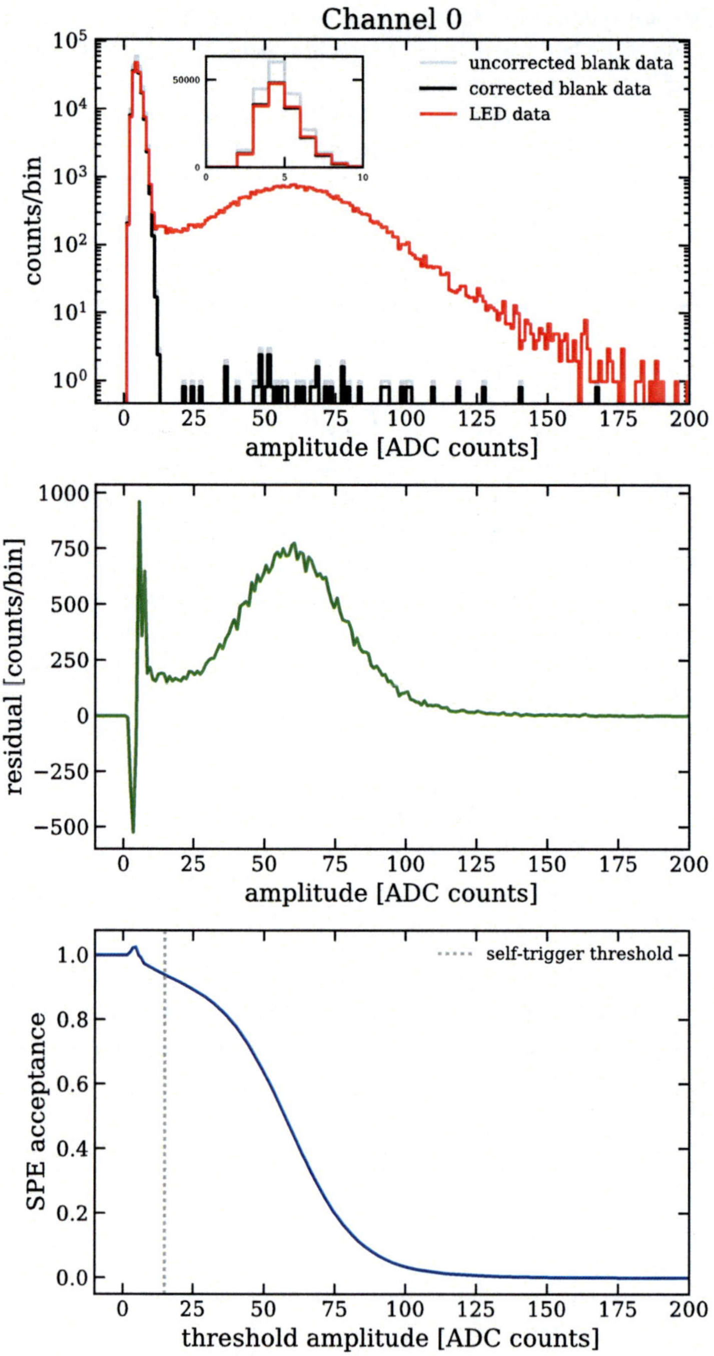

Fig. 2.10 Example outlining the procedure to extract the SPE acceptance. See text for details

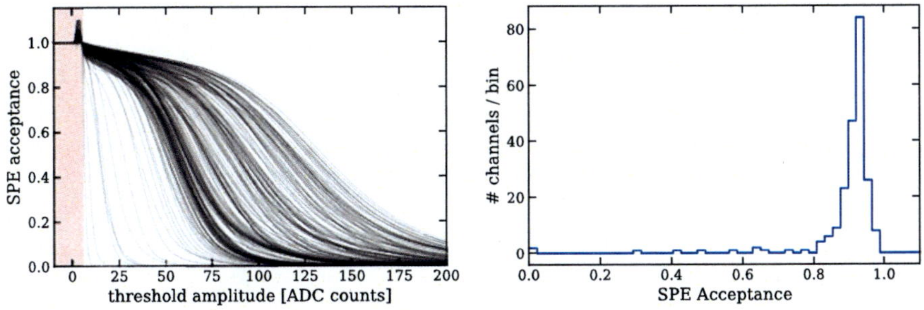

Fig. 2.11 Left: SPE survival functions of all active PMTs. Right: SPE acceptance values for all active PMTs in a single calibration run of XENON1T

Since we are interested in the SPE acceptance, we would ideally have the 2^{+}PE component be negligible; however, such occupancies would also make it difficult to have enough SPE statistics, and so we have to find a balance. For the example shown in Fig. 2.10, $\lambda = 0.22$, implying that with $P(k \geq 2) = 0.02$ the 2^{+}PE component is sufficiently small to be ignored. The average occupancy across all PMT channels is $\overline{\lambda} \approx 0.17$ giving an average 2^{+}PE of $\sim$1%. We thus conclude that the occupancies are small enough that the contribution from 2^{+}PE can be neglected.

After correcting the blank spectrum for the occupancy and ignoring the 2^{+}PE contribution, we can extract the SPE amplitude distribution by doing a statistical subtraction of the (corrected) blank spectrum from the LED one. This is shown in Fig. 2.10 (middle), where the spectrum is observed to be roughly Gaussian but deviates at low amplitudes. This is due to *under-amplified* photoelectrons, signals that are detected but not amplified by the full dynode chain of the PMT and so display a smaller amplitude. A systematic effect is seen near the threshold $t = 6$ due to the correction we make on the blank dataset. Since this artifact is integrated to determine the acceptance, this is the largest source of uncertainty in this study. From this SPE amplitude spectrum, we calculate the survival distribution, shown in Fig. 2.10 (bottom), which gives the fraction of the SPE distribution greater than or equal to a given amplitude. This survival distribution gives us the SPE acceptance as a function of self-trigger threshold; for this run and channel the threshold was set to 15 ADC counts, implying an SPE acceptance of $\sim$94%, which is also shown in Fig. 2.10 (bottom).

We repeat this procedure for all PMT channels, the results of which are summarized for a single calibration run in Fig. 2.11. The average SPE acceptance across all channels for this dataset was 89.6%. The outliers in this plot are PMTs that have low gain and poor SPE acceptance, with PMTs of <50% acceptance ignored in all XENON1T analyses. We then tracked this average acceptance as a function of time, which is shown in Fig. 2.12, along with the mean and 1- and 2-σ bands. The time dependence shows that the acceptance was mostly stable; however, there are a handful of outliers not completely understood. The errorbars include statistical and systematic uncertainties, the latter of which is dominated by the choice of t,

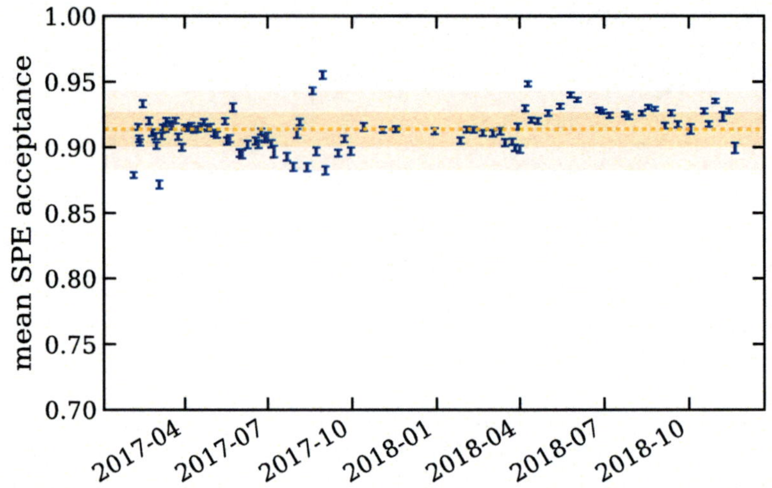

Fig. 2.12 The average SPE acceptance across all active PMTs in XENON1T, as a function of time

which we vary between 5 and 8. We note that it is possible (and perhaps likely) that the errorbars are underestimated, as it is difficult to quantify every systematic effect in this analysis, especially since it assumes the noise conditions are identical in two different datasets. We conclude that our SPE acceptance is stable within the 90–95% range for the time region considered here, which as we will see is precise enough for this analysis.

As mentioned before, the SPE acceptance is related to the detection efficiency via the term $P(n_h|n_{phd})$. The SPE acceptance gives the probability that a single detected photon would induce a hit that was detected and stored. Therefore, $P(n_h|n_{phd})$ can be modeled as a binomial distribution with p given by the SPE acceptance. In reality, each channel has its own SPE acceptance and so the detection efficiency would have a non-trivial spatial dependence; however, this effect is surely small, so we take an average SPE acceptance p for all PMTs. Given this, the probability of detecting n_h is given by the binomial distribution:

$$P(n_h|n_{phd}, p) = \binom{n_{phd}}{n_h} p^{n_h}(1-p)^{n_h - n_{phd}} \tag{2.13}$$

From this section we know that $p \approx 0.9$, and so it is straightforward to calculate this component to the overall efficiency for a given number of detected photons.

2.2.2 *Reconstruction Efficiency*

We discuss here the efficiency of the XENON1T data reconstruction algorithms. TPC signals that exceed the self-trigger threshold of PMTs are processed using software called PAX, for Processor for Analyzing XENON [2]. This software clusters individual PMT hits into *peaks*, which are then classified as S1 or S2 signals. We study here the efficiency of PAX at reconstructing and classifying true S1 signals, which is the dominant component to the overall S1 efficiency.

In XENON1T we developed two methods to estimate the S1 efficiency.

1. The first is a pure simulation-based method, where we simulate photons and electrons in the TPC, distribute their signals across individual PMTs, simulate the PMT charge response to create waveforms, and then pass those waveforms to PAX for reconstruction. We refer to this as the FAX method.
2. The second is a purely data-driven method and is what we focus on in this work. In the data-driven method, also called the bootstrapping method, we randomly sample individual hits from large S1 waveforms (where the reconstruction efficiency is assumed to be unity), to create fake, low-energy signals that are then passed to PAX for reconstruction.

The data-driven method works as follows. Hits from large S1s ($\mathcal{O}$(50–100) PE) in ^{220}Rn calibration data were random sampled to create fake, low-energy events. The idea is that, by sampling from a true S1, the true pdf of the waveform shape is sampled in a data-driven way. We refer to the large, real S1s as the *parent* events and the bootstrapped, fake S1s as the *child* events. These children were then passed through PAX to see whether it properly reconstructs them. This analysis uses PAX v6.10.1, where the peak classification algorithm is based on three variables:

1. *tight coincidence*, the number of channels that observed a hit within 50 ns of the maximum amplitude of the peak.
2. *rise time*, defined to be the time difference between the points where the signal reaches 10 and 50% of the total area.
3. *width*, defined to be the time difference between the points where the signal reaches 5 and 95% of the total area.

In order for a peak to be classified as an S1, the PAX reconstruction software requires that *tight coincidence* ≥ 3 (or 2[1]), *rise time* < 70 ns, and *width* < 300 ns.

To validate the data-driven method of estimating the efficiency, we studied whether the child events actually look like real data. We did this by comparing the tight coincidence, rise time and width variables of fake data (from FAX and data driven method) to *real* low energy data acquired during ^{37}Ar calibration.[2] For the simulated events, we consider reconstructed but pre-classified peaks; i.e., the parameters are calculated for the peaks but no S1 or S2 classification has been performed. For the real data we consider the fully reconstructed S1 data, where efficiency loss will make the distributions differ. Due to this fundamental difference between the real and simulated/bootstrapped data, we are primarily interested in

[1] Our analysis that follows in the later chapters will use a threefold coincidence requirement, but we still calculate to the twofold one here.

[2] For more details on ^{37}Ar, see Sect. 2.1.1.

qualitative statements about the validity of the fake waveforms, and then quantitative estimates of the efficiency will come afterwards.

Comparisons of the data-driven and FAX methods to real ^{37}Ar data are shown in Fig. 2.13. To make sure we are not biasing the analysis based on underlying energy spectra, we plot the time profiles (rise time and width) for each type of data for a given tight coincidence value. Since we are interested in efficiency loss near the energy threshold, we only show comparisons for tight coincidences 2, 3, 4, and 5, but consider higher values in the efficiency analysis. We observe a good match between the simulated data (both data-driven and fax methods) and the real data, apart from some differences in the tails as expected due to the reconstruction efficiency being already applied to the real data. It appears that both the data-driven and FAX methods succeed in modeling the time profiles of real S1 waveforms. From this check we conclude that the two methods are valid and that we can use them both to extract the reconstruction efficiency.

After their validation, the child S1 events are passed to the PAX reconstruction software to see how often it correctly reconstructs them. There are four possible outcomes that we consider:

1. The S1 is found and classified correctly.
2. A peak was found, but it was left *unclassified*. This happens for peaks that satisfy neither the requirements for an S1 nor an S2, which almost always is due to the tight coincidence threshold.
3. A peak was found, but it was classified as an S2 instead of an S1
4. No peak was found, not even a lone hit. This is expected to happen very rarely.

The results are shown in Fig. 2.14, where the outcome fractions are given as a function of number of true hits. This plot uses a threefold tight coincidence requirement, but we also performed the analysis for a twofold one. For threefold coincidence requirement, there is no efficiency for $n_h < 3$, with all of the S1s being unclassified. The S1 efficiency never reaches 100% due to ~5–10% of events being mis-classified as S2 signals for $n_h > 6$. Similar results hold for the twofold efficiency, but with higher 'found' efficiency and lower 'unclassified' at low numbers of hits due to the lower coincidence threshold. Since we are primarily interested in $\epsilon_{\text{rec}}(n_h)$, the efficiency of correctly finding the S1s, we summarize the results for both two- and threefold coincidence in Fig. 2.15.

Figure 2.15 shows the efficiency in the space of *hits detected*, not *photons detected* as in Eq. (2.9). To determine $\mathrm{P}(S1|n_{phd})$, we must account for the SPE acceptance. From Sect. 2.2.1, particularly Eq. (2.13), we can calculate ϵ_{det} for different values of SPE acceptance p. From Fig. 2.12, we have (1-σ band)

$$p = 0.914 \pm 0.013 \tag{2.14}$$

In Fig. 2.16 we show the final S1 detection efficiency after convolving Fig. 2.15 with the SPE acceptance ala Eq. (2.9). We consider two sources of uncertainty (in quadrature), the propagated statistical uncertainty and the 1-σ systematic uncertainty from the SPE acceptance.

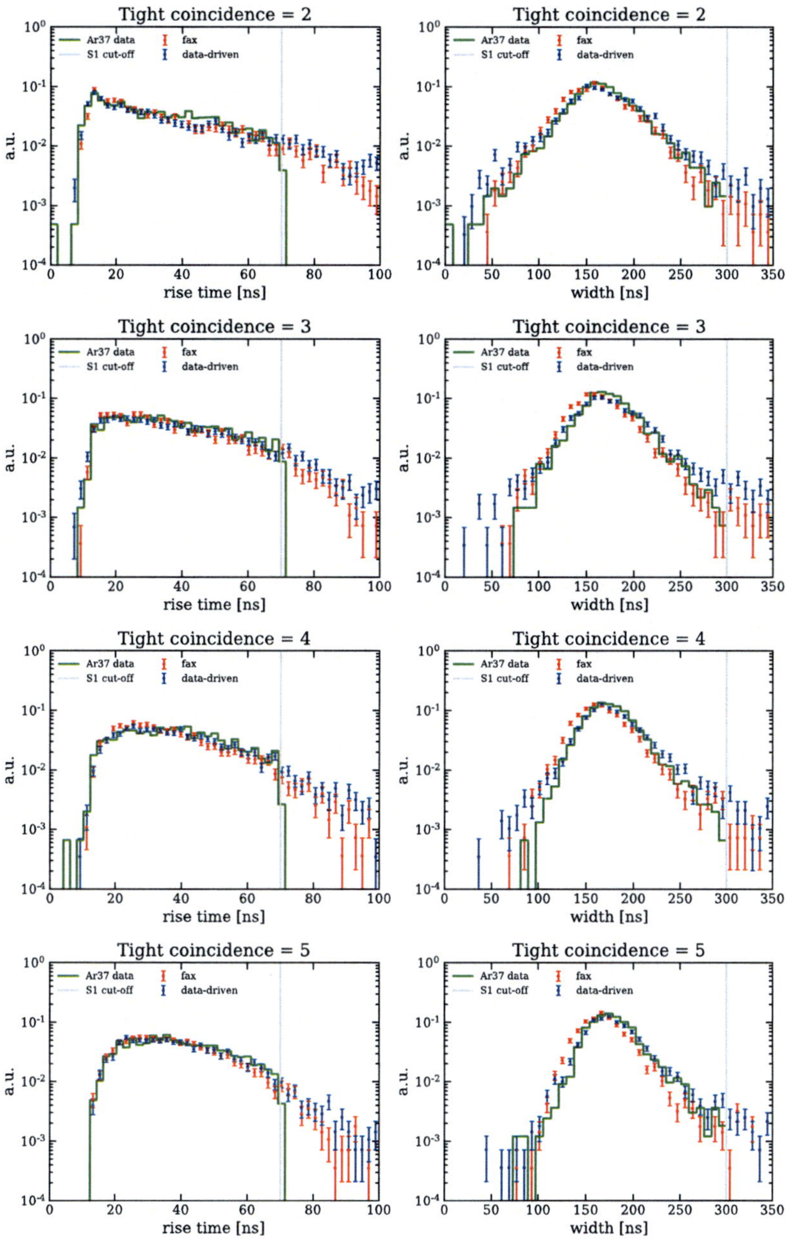

Fig. 2.13 Comparing the time profile of simulated waveforms from the data-driven (navy) and FAX (red) methods to real data from ^{37}Ar calibration (green). Gray lines show the thresholds required by PAX to be classified as an S1. Note that the ^{37}Ar points have been reconstructed with PAX already, so there is efficiency already applied. Considering this, we think the match between real data and the simulation/bootstrapping methods is good enough to trust their results for the reconstruction efficiency

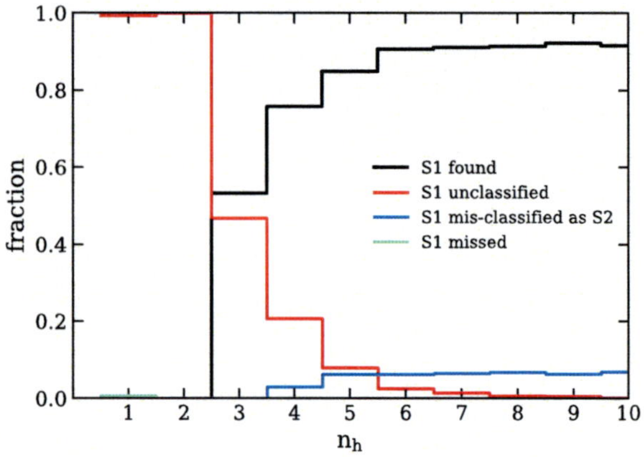

Fig. 2.14 The fractional outcomes for the reconstruction efficiency, as determined using the data-driven method. This assumes a threefold tight coincidence requirement

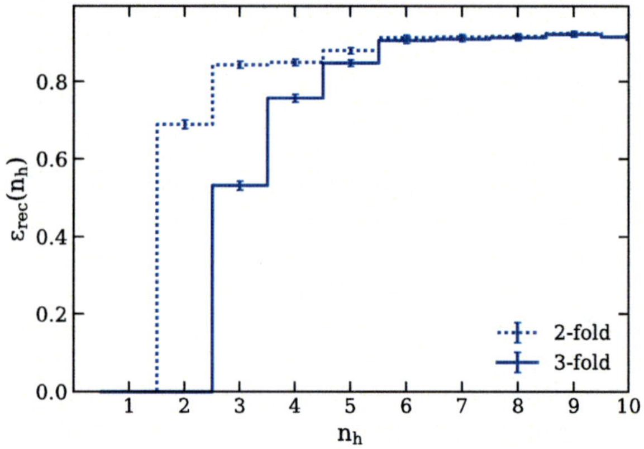

Fig. 2.15 The total reconstruction efficiency as a function of number of PMT hits n_h, as extracted using the data-driven method. Efficiency curves for both two- and threefold tight coincidence are included. This figure considers only the efficiency of the data processor PAX at finding and classifying S1s in the raw data, not the *detection* efficiency loss due to the non-unity SPE acceptance

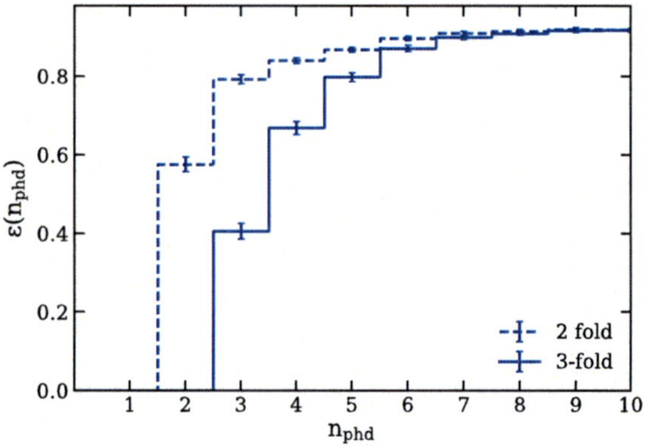

Fig. 2.16 The total S1 efficiency as a function of photons detected, extracted using the data-driven method. Efficiency curves are included for both two- and threefold tight coincidence thresholds

We now compare the data-driven efficiency in Fig. 2.16 to that derived from FAX. There are many details of the FAX simulation that we are glossing over, but the extraction of the efficiency is straightforward: we know the number of photons

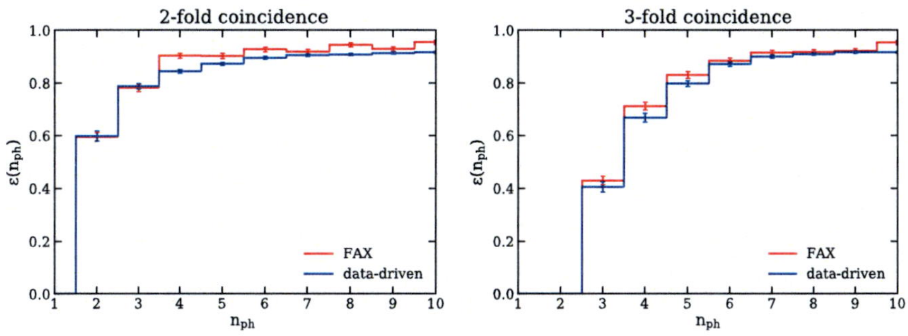

Fig. 2.17 Comparison between FAX and the data-driven method outlined here, for both twofold (left) and threefold (right) coincidence

detected, and we determine how often we reconstructed and classified an S1 on the other side of the simulation + processor. Comparisons for both twofold and threefold coincidence thresholds are shown in Fig. 2.17, where overall good agreement is observed. The data-driven method suggests slightly worse efficiency than that given by the FAX method. This could be due to effect *not* modeled in FAX, or it could be a systematic effect not accounted for in the data-driven analysis. For the final efficiency we thus use the mean of the two analyses, with the differences being included as a systematic uncertainty.

2.2.3 Efficiency as Function of Charge

Above we consider the total S1 efficiency in the space of photons detected. In reality, we want the efficiency in the space used for our analysis: the reconstructed energy, which is related to the integrated charge collected on each PMT. In this section, we convert the efficiency in Fig. 2.17 first to the space of charge, and finally to the space of the energy.

To convert from n_{phd} to the space of charge c, we can expand the efficiency as

$$\epsilon_c(c) = \sum_{n_{phd}} \epsilon(n_{phd}) f_c(c|n_{phd}), \tag{2.15}$$

where f_c is the probability density function of the observed charge for a given number of detected photons. To get f_c we first need the SPE charge distribution, which we extract using a very similar method (and the same calibration data) as in Fig. 2.10, where we extract the *charge* spectrum rather than the amplitude one. We smear over the variation of each PMT by normalizing the charge distribution to have a mean of 1 PE, and then taking an average over all active channels to get an average

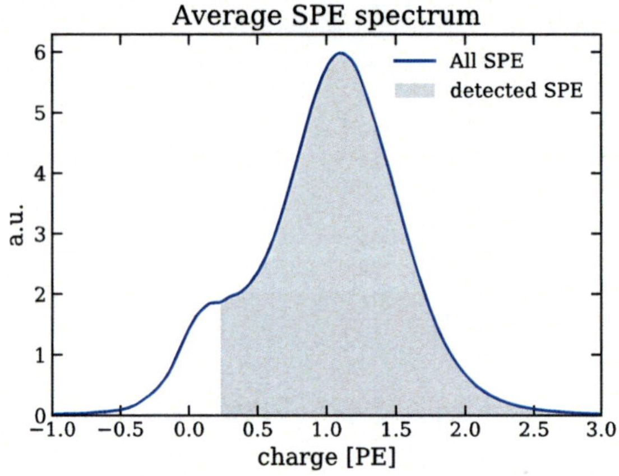

Fig. 2.18 Average SPE charge distribution (convolved with noise), in units of photoelectrons [PE]. Since we know the SPE acceptance is ≈90%, we assume that the 10th percentile of the distribution is not detected, leaving the largest 90% (gray) as the detected SPE signals

SPE charge distribution, shown in Fig. 2.18. Given our measured SPE acceptance, we assume that the largest ∼90% of the distribution is what we actually detect.

We also need to account for an effect known as *double photoelectron (DPE) emission*, wherein a single UV photon from xenon scintillation actually produces two photoelectrons instead of one. This effect occurs much less frequently with visible light, so the LED data cannot be used to calibrate it. For the model of PMTs used in XENON1T, it has been measured that the DPE probability ranges from ∼0.18–0.24 [91].

To convert from photons to charge space, we use a toy Monte Carlo (MC) technique. For a given number of photons, we simulate the effect of SPE acceptance with a Bernoulli/binomial distribution, after which the 'detected' signals are assigned charge values by sampling from the gray region in Fig. 2.18 (which changes slightly depending on SPE acceptance), sometimes sampling two values according to the given DPE probability. This simulation gives the distribution of charge vs. number of hits, which we can then compare to real data. Assuming an SPE acceptance of ∼0.9 we find the best agreement to ^{37}Ar calibration data for a DPE probability of 0.19, a value consistent with dedicated measurements. Comparisons of the toy MC and real data distributions are shown in Fig. 2.19, including a toy MC for DPE=0.19 and DPE=0. The DPE=0.19 toy distribution does not match perfectly the shape of the real distribution, but the means of the distributions agree at the level of ∼1% for all $n_h < 20$ (and it agrees much better than, e.g., DPE=0, as shown). The different shapes of the distributions can likely be attributed to the fact we smear out PMT-specific behavior by taking an average SPE acceptance and an SPE charge distribution; this would lead to a worse charge resolution, consistent with the differences seen in Fig. 2.19. Including the information for each PMT in the simulation would be a nice expansion of this study, but this is not done here.

From the simulations described above, with examples shown in Fig. 2.19, we can calculate the f_c term in Eq. (2.15). As systematic checks, we vary the SPE acceptance within 1-σ, and compare to the FAX method for estimated the efficiency.

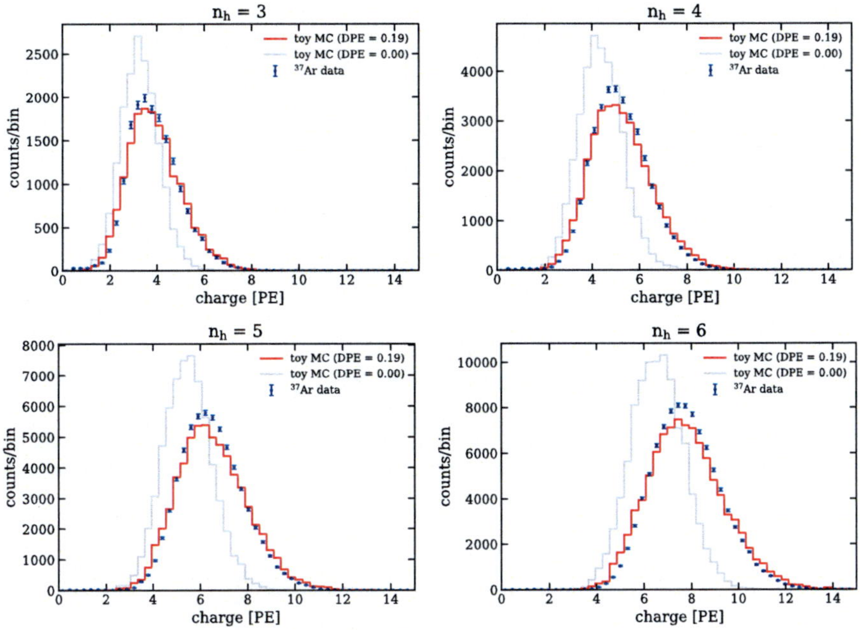

Fig. 2.19 S1 charge distribution in slices of n_h, comparing real data to a toy simulation described in the text. This study suggests a DPE fraction of 0.19

The final charge efficiency for both two- and threefold coincidence is shown in Fig. 2.20.

2.2.4 Efficiency as Function of Reconstructed Energy

As the final step, we convert the efficiency to the space of reconstructed energy. Similarly to the charge case, we can expand

$$\epsilon_E(c) = \int \epsilon_c(c) f_E(E|c) dc, \tag{2.16}$$

where E is the reconstructed energy and f_E the pdf for E given a known value of c. To get f_E, we use the XENON microphysics model, colloquially known as "bbf" (for Bayesian Band Fitting). This model was used to fit the full response of XENON1T, the posterior of which can be used to model signals of any kind in the detector. Details about bbf can be found in [36].

The bbf model allows for simulation of each step of the detection process, taking as input deposited energy and event position and returning distributions of S1 and S2, in the process accounting for microphysics processes as well as

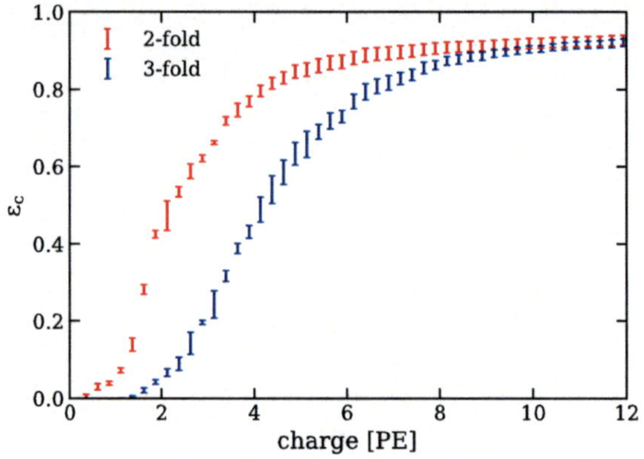

Fig. 2.20 Total S1 efficiency in the space of detected charge, for both two- and threefold tight coincidence. Uncertainties are dominated by systematic effects of comparing the data-driven and FAX methods and varying the SPE acceptance

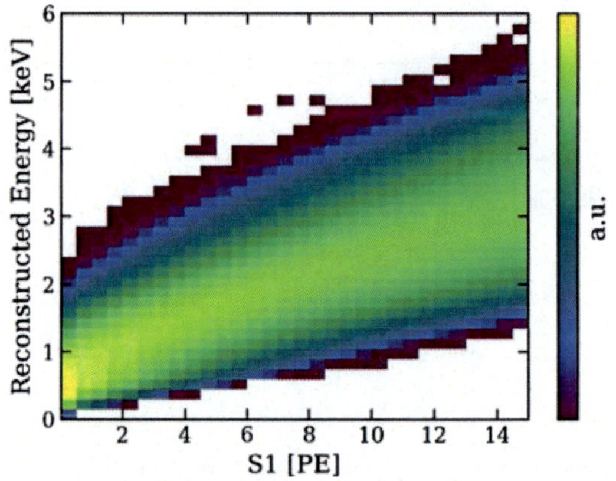

Fig. 2.21 Distribution of reconstructed energy E_r as a function of S1, as determined using the XENON microphysics model called bbf. Convolution of this distribution with the S1 efficiency in Fig. 2.20 allows for conversion of the efficiency to the space of reconstructed energy

position-dependent detector effects described in Sect. 2.1.1. It is thus a useful tool for converting the efficiency in S1 space (Fig. 2.20) to the space of reconstructed energy. To do this, we simulated electronic recoils uniformly distributed in the TPC and with energies between 1 and 10 keV. While the bbf simulation itself does include efficiency loss, this was turned off in the simulation for the purposes of this study. Since this simulation includes the position information, it also allows for calculating for each simulated event the $cS1$ and $cS2_b$ values, and, using the g_1 and g_2 from Sect. 2.1.1, hence the reconstructed energy E_r. At the end of this simulation, we thus can determine the distribution E_r as a function of (uncorrected) $S1$. This distribution is shown in Fig. 2.21

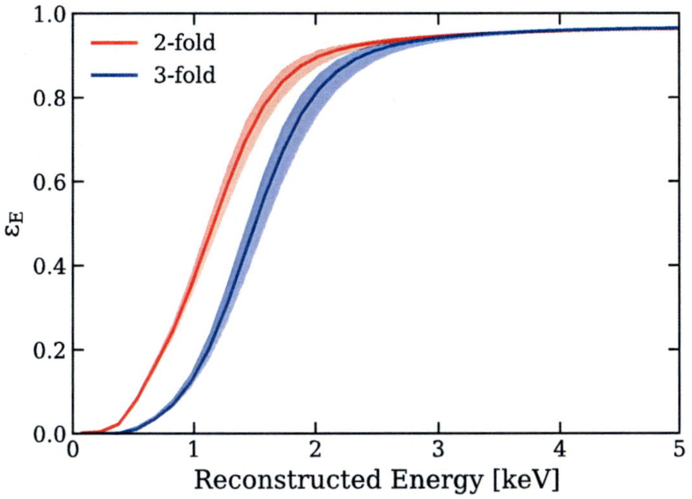

Fig. 2.22 Total S1 efficiency in the space of reconstructed energy, for both two- and threefold tight coincidence. Uncertainties include systematic effects of comparing the data-driven and FAX methods, varying the SPE acceptance, and varying the photon yield parameter in bbf

The bbf simulation also allows for varying, e.g., microphysics parameters, giving slightly different versions of Fig. 2.21. As a study of systematics, we varied several parameters of the bbf posterior within their 1-σ values, where we found that the photon yield was the most dominant contribution. We include this as an additional systematic uncertainty in the total conversion.

After convolving Fig. 2.20 with Fig. 2.21, the efficiency in the space of E_r can be determined. The resulting efficiency curves in the space of reconstructed energy are shown in Fig. 2.22, for both two- and threefold tight coincidence values. These are the main efficiency curves that will be included in the search for new physics in the electronic recoil channel with XENON1T. Uncertainties include: statistical uncertainty from the data-driven method, systematic uncertainty from the difference between the data-driven method and FAX, systematic uncertainty from the SPE acceptance, and systematic uncertainty from the microphysics parameters used in the bbf simulation.

The final model for the energy threshold in this analysis is summarized by Fig. 2.22. To validate this model, we consider ^{220}Rn calibration data, whose spectrum at low energies is dominated by the β-decay of ^{212}Pb. We use a theoretical model for the spectrum of ^{212}Pb that includes atomic effects [40], smear the true spectrum according to (2.8), and then apply the efficiency to get the predicted spectrum in the space of reconstructed energy. This prediction is compared to the real data (with selection cuts applied, see Sect. 2.3), as shown in Fig. 2.23. For both 2- and threefold coincidence we observe good agreement between the model and the data, with χ^2/d.o.f. = 0.93 ($p = 0.59$) for the twofold comparison and χ^2/d.o.f. = 1.40 ($p = 0.06$) for the threefold. If we set a 1-keV energy threshold—as we do in

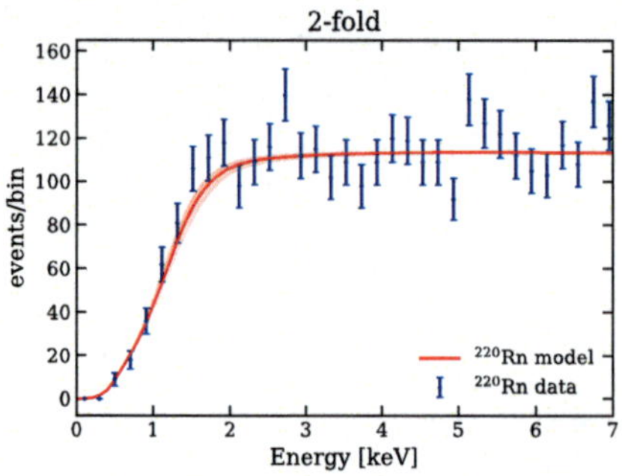

Fig. 2.23 Comparison between ^{220}Rn calibration data and a model based on the total S1 efficiency calculated in this chapter. The good agreement, especially above 1 keV, validates the efficiency calculation performed here

the main analysis described in succeeding chapters—the agreement is even better, with χ^2/d.o.f. = 1.05 (p = 0.39) and χ^2/d.o.f. = 1.15 (p = 0.26) for two- and threefold, respectively. The improvement in the fit is due to a small disagreement between the data and the model at very low energies where the efficiency is quite poor. These χ^2 values do not include the uncertainty on the efficiency itself; only the statistical uncertainty on the data points is considered. A similar study using ^{220}Rn calibration data is described in Sect. 4.2.

To conclude, we have calculated the total S1 efficiency using a data-driven method based on (1) calibration of the SPE acceptance and (2) simulated waveforms via bootstrapping large S1 signals. This S1 efficiency was then converted to the space of reconstructed energy using the XENON microphysics model known as bbf, the result of which shows good agreement with the observed spectrum from ^{220}Rn calibration data. This efficiency is an important input in the background- and signal-modeling described in later chapters.

2.3 Event Selection

In this section we briefly summarize the event selection for this analysis, which is quite similar to [31].

Run Selection As foreshadowed previously, we focus on a dataset called SR1 that spanned from February 2017 to February 2018. We also include a dataset called SR2, which in full lasted from February 2018 to December 2018, but, as we discuss in more detail in Sect. 4.8, very little of this exposure is useful for this analysis. Due to time-dependent backgrounds on account of online krypton distillation, we do not consider SR0, which was used for the first XENON1T results [26]. Additionally, events within 24 h from the end of calibration campaigns that used injected radioactive sources were removed due to residual source activity.

Fiducial Volume This analysis is performed with a cylindrical fiducial volume, with the bottom ~7 cm, top ~9 cm, and the outer ~13 cm of the TPC removed from consideration. The corresponding total xenon mass is 1042 kg. This fiducial volume is smaller than, e.g. [31], but is used to reduce the *surface* and *material* backgrounds to a ~ negligible level, which simplifies the analysis significantly. These backgrounds, discussed in more detail in Sect. 3.1, have a strong spatial dependence with events localized to the outer edges of the detector.

Energy ROI Events with energies between (1, 210) keV are selected for this search. The 1-keV bound was determined by requiring the total efficiency be larger than 10%, shown in Fig. 2.22 (right), and the 210-keV bound was chosen due a large γ−ray background from detector materials, which is difficult to model due to large uncertainties on its spectral shape, above this energy.

Single Scatter Cuts Since events from the signals described in Chap. 1 are exceedingly rare, events with multiple interaction sites are removed. There are three cuts used for this purpose:

1. For events with multiple S1s, the second-largest S1 is checked for compatibility with the S2, based on the two signals sizes, the width of the S2, and the time difference between them. If the second S1 is compatible with the S2, the event is removed.
2. For events with multiple S2s, if the second-largest S2 is too large compared to the main S2, the event is removed.
3. A cut designed specifically for the ^{83m}Kr background (see Sect. 3.1), where rarely one of the two S1 signals was mis-classified as an S2. These events are removed, so that only events with the unresolved, 41.5-keV signal from this source was included.

S1 Threshold We use a threefold coincidence requirement for detection of S1s, meaning we will be using the corresponding threefold efficiency curve calculated in the previous section. While a twofold coincidence implies a better efficiency, it also introduces a larger background from *accidental coincidence* (AC) events, where uncorrelated S1-like hits (usually dark counts) randomly occur near the same time and are paired together. See Sect. 3.1 for more details on the AC background.

S2 Threshold To suppress the AC and surface backgrounds to a further negligible level [36], an S2 threshold of 500 PE is applied. This does not remove any ER events, but does remove these instrumental backgrounds. A 500 PE S2 threshold is also large enough to ensure 100% detection efficiency on the S2 signal, simplifying the efficiency study from Sect. 2.2.

Data Quality Cuts A variety of other selection cuts are applied to ensure data quality and a correct S1 and S2 pairing, as detailed in [37]. Briefly:

1. A cut that removes events where the DAQ was in a busy state.
2. A cut that removes events that follow too closely in time to large S2 signals, where the tails of the S2 signal can bias the event reconstruction.

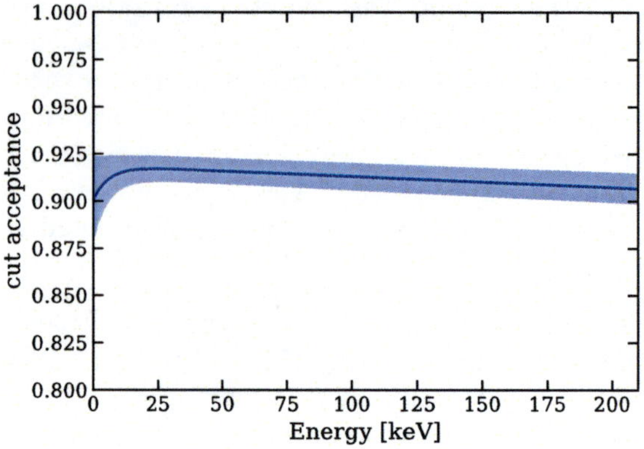

Fig. 2.24 Summary of the acceptance of all selection cuts applied to SR1 data. The acceptance is quite flat with energy, with a small dip at low energies of $\lesssim 1\%$

3. A cut that removes events in rare periods of unstable PMT behavior, such as when PMTs themselves are emitting light.
4. XENON1T uses two different position reconstruction algorithms, a FANN and top-pattern likelihood fit (see Chap. 1). If the difference in event position is large for these two, the event is removed.
5. Due to diffusion, the width of an S2 depends on its depth in the detector. A model for this was developed, where the width of an S2 could be predicted using the S2 size and the drift time [1]. Events in which the observed S2 width does not agree with the model are removed.
6. Events with unphysical PMT hit patterns for either the S1 or the S2 are removed.
7. Events where an unphysical amount of light is seen in the top PMT array for an S1 or an S2 are removed.
8. Events where one PMT is contributing to a large fraction of the total charge are removed.

Final Cut Acceptance The efficiencies and uncertainties of the selection cuts are estimated analytically when possible; otherwise they are estimated using ^{220}Rn calibration data [37]. The cumulative selection efficiency after applying all cuts was fitted using empirical function, from which an average cumulative selection efficiency over the $(1, 210)$ keV region of $(91.2\pm0.3)\%$ is determined. A summary of the empirical fit and uncertainties is shown in Fig. 2.24. For more details, see [163].

Summary of Event Selection The final effective SR1 live time is 226.9 days and thus the total exposure for this analysis is 0.65 tonne-years. In the next chapter, we build a background model to predict the energy spectrum for this exposure within the $(1, 210)$ keV ROI, and develop the statistical framework used to search for signs of new physics with this SR1 dataset.

Chapter 3
Background + Signal Modeling and Statistical Methods

Abstract In the previous chapter we discussed how the energy of an ER event is reconstructed, along with the detection efficiency and event selection. Here we discuss the sources of background present in the SR1 dataset, having survived the event selection cuts described the previous chapter. We use the inputs from Chap. 2 regarding the efficiency and resolution to construct models for the energy spectra of all backgrounds and possible signals presented in Chap. 1. Armed with these background and signal models we can then search for signs of new physics using a statistical analysis, the details of which are discussed at the end of this chapter. We report the results of the statistical inference on the real XENON1T data in Chap. 4.

3.1 Background Model

Within the $(1, 210)$ keV ROI and the 1042 kg fiducial volume we identify ten sources of background, summarized in Table 3.1. Six components exhibit continuous energy spectra and are modeled based on theoretical predictions and/or GEANT4 Monte Carlo simulations, and the rest are mono-energetic peaks that are modeled as Gaussian functions of known energies and resolution. Instrumental backgrounds like accidental coincidence and surface backgrounds are negligible due to our event selection. For all backgrounds, the true energy spectra are smeared according to the detector resolution and the expected efficiency loss accounted for. The rates of the background components are constrained, when possible, by independent measurements and extracted by the fit.

The true energy spectra of all background and signal models are converted to the space of reconstructed energy using Eq. (2.8), smearing with a Gaussian resolution function with width given by the empirical function in Fig. 2.8. We note that this Gaussian smearing is an approximation; at low energies the response of the detector is not Gaussian, and thus we are introducing a systematic uncertainty using this method. However, the dominant source reconstruction bias at low energies is due to the efficiency, which we do account for [40].

After smearing, the efficiency from Fig. 2.22 (three-fold) is applied to the spectrum to account for the energy threshold, and the cut acceptance from Fig. 2.24

E. Shockley, *Study of Excess Electronic Recoil Events in XENON1T*,
Springer Theses, https://doi.org/10.1007/978-3-030-87752-1_3

Table 3.1 Summary of background components

No.	Component	Origin	Spectral type	Time dependence in SR1
1	^{214}Pb	Internal contaminant	Continuous	Constant
2	^{85}Kr	Internal contaminant	Continuous	Slight increase
3	^{136}Xe	Intrinsic	Continuous	Constant
4	^{83m}Kr	Internal contaminant	Peak	Decay
5	^{133}Xe	Neutron activation	Continuous	Activation and decay
6	^{131m}Xe	Neutron activation	Peak	Activation and decay
7	^{124}Xe	Intrinsic	3 peaks	Constant
8	^{125}I	Neutron activation	3 peaks	Activation and decay
9	Materials	External contaminant	Continuous	Constant
10	Solar ν	Astrophysics	Continuous	Subtle modulation (ignored)

is applied. In this way, the true spectra of background and signal models are converted to the expected observed spectra in the space of reconstructed energy. These models can then be directly compared to the energy spectrum of the data, which is determined using the methods described in Sect. 2.1.1. No further corrections need to be applied to the data spectrum.

As we discuss in more detail below, five of the background components exhibit time dependence throughout the SR1 dataset, which we account for by partitioning the 227-day exposure into two sub-datasets. Thus we will include both spectral and temporal models for each component in the discussion that follows.

Below we discuss each of the 10 background components in detail.

^{214}Pb The β-decay of ^{214}Pb, the dominant continuous background at low energies, is present due to ^{222}Rn emanation into the LXe volume by materials. While mitigation techniques through material screening and distillation have come a long way, small amounts of radon outgasses from detector components into the xenon and, as a noble gas, is not removed by the xenon purification system. This radon then dissolves in the liquid xenon medium, where it can decay down the chain, eventually to ^{206}Pb. First, ^{222}Rn undergoes α-decay to ^{218}Po with $T_{1/2}$=3.8 days. With a half-life of O(minutes) ^{218}Po itself α-decays to ^{214}Pb, the background in question, which undergoes β-decay to ^{214}Bi with a half-life of $\sim$30 min. About 90% of these β-decays go to an excited state of ^{214}Bi, which immediately (O(ns)) deexcites by emitting a γ with E $\geq$ 295 keV in addition to the original β. These multi-step decays may be reconstructed as (1) unresolved, single event with E $\geq$ 295 and thus outside our (1, 210) keV ROI, or (2) resolved, multi-scatter event which would be removed by the single-scatter selection cuts described in Sect. 2.3. The other $\sim$10% of ^{214}Pb β-decays go directly to the ^{214}Bi ground state and so include no such γ; these are the decays relevant for this analysis. Refer to [162] for details about the specific decays discussed above.

We can place a lower limit on the ^{214}Pb activity by constraining the activity of its daughters. Continuing down the decay chain, ^{214}Bi decays (β, $T_{1/2} \approx 20$ min) to ^{214}Po, which then rapidly ($O(\mu s)$) undergoes α-decay to the effectively-stable ^{210}Pb. Since XENON1T can resolve decays separated by ~μs , these so-called ^{214}BiPo events are easy to pick out by looking for the coincident β + α. It follows that, using the α-decays of ^{218}Po (parent of ^{214}Pb) and the ^{214}BiPo decays (daughter of ^{214}Pb), we can place upper and lower limits on the activity of ^{214}Pb. The results of these studies are summarized in [36], where we constrain the ^{214}Pb activity to be in $(5.1 \pm 0.5, 12.6 \pm 0.8)$ μBq/kg, assuming a uniform distribution in the liquid xenon.

For the shape of the true ^{214}Pb spectrum, we use a calculation from X. Mougeot [40]. It is known that β-spectra can be affected, especially at low energies, by atomic screening and exchange effects as well as nuclear structure [62, 108]. The β decay of ^{214}Pb is a first forbidden non-unique transition; however, spectra from IAEA LiveChart (Nuclear Data Services database) [154] are based on calculations assuming an allowed transitions, and further do not include exchange effects or screening effects that can affect the low-energy region of the spectrum [130]. The ^{214}Pb model from GEANT4 [6] does at least include the screening effect, but its implementation displays a non-physical discontinuity at low energies [111, 130]. Accordingly, a dedicated calculation was made for this work to account for possible low-energy discrepancies from these effects. Such calculations were made for the ^{214}Pb, ^{85}Kr and ^{212}Pb spectra, the last of which is important for calibration data. This theoretical spectrum was then converted to the space of reconstructed energy, as discussed above, giving the final ^{214}Pb background model shown in Fig. 3.1 (left).

The time dependence of ^{214}Pb was determined by monitoring the rate of the correlated α-decays, similar to the constraints on the overall rate. This was done for both SR1 and SR2 using the ^{218}Po αs, with the results shown in Fig. 3.1 (right). The rate is very stable during SR1, but displays a strong temporal behavior during SR2 due to improvements in the xenon purification system and a radon distillation campaign, both of which reduced the radon concentration in the detector. Tests of the purification system also led to small, step-wise increases in the radon concentration. An empirical model was then developed, based on known purification

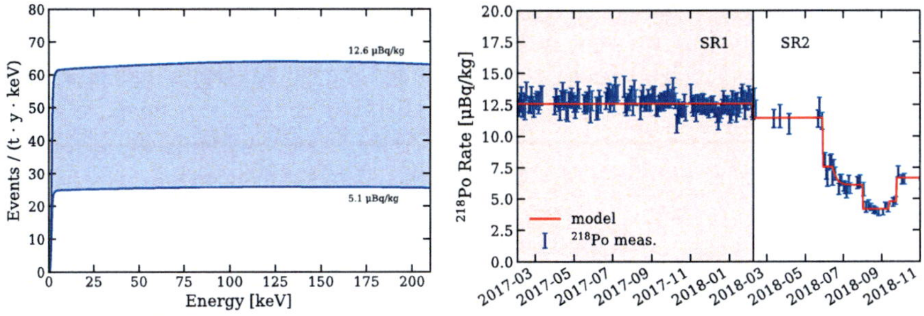

Fig. 3.1 Background model for ^{214}Pb in energy (left) and time (right) for both SR1 and SR2. Both models were constructed using α-decays in the ^{222}Rn chain

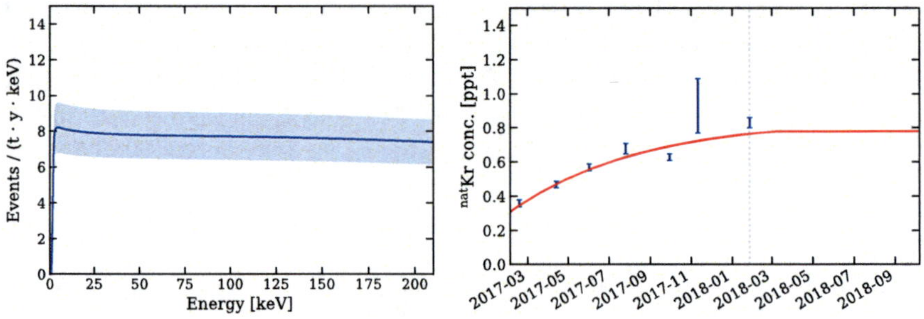

Fig. 3.2 Background model for ^{85}Kr in energy (left) and time (right) for SR1. Both models were constructed using dedicated RGMS measurements of the krypton concentration in the detector. The band in the left figure represent the 1-σ uncertainty, dominated by systematic uncertainty on the RGMS measurements [124]

improvements/testing, and fitted through the data points [38]. This model is also shown in Fig. 3.1 (right), the results of which is used for the time-evolution model of ^{214}Pb.

^{85}Kr An additional background comes from intrinsic ^{85}Kr, which is subdominant due to its removal via cryogenic distillation [28, 131]. Similarly to ^{214}Pb, this is a β emitter, with the true energy spectrum including atomic exchange and screening effects calculated for this work in [40]. The final model in reconstructed energy is shown in Fig. 3.2 (left), with the decay rate was inferred from dedicated measurements of the isotopic abundance of ^{85}Kr/^{nat}Kr (2×10^{-11} mol/mol) and the ^{nat}Kr concentration evolution in LXe [124]. These measurements give a weighted average of 0.66 ± 0.11 ppt during SR1, which was converted to an event rate using a scale factor of 11.64 events/(t·y·keV)/ppt derived using early-XENON1T data (before distillation), when the low-energy background rate was dominated by ^{85}Kr. The uncertainty on this measurement is dominated by the systematic uncertainty on the RGMS measurements of the ^{nat}Kr concentration and is shown as the shaded band in Fig. 3.2 (left).

The same measurements also allow for the time-dependence of the ^{85}Kr decay rate to be taken into account, as shown in Fig. 3.2 (right). An increase of < 1 ppt/year is observed. This could be due to outgassing from detector materials, which is expected after the conclusion of krytpton distillation, or from a small air leak. At this time there are no measurements of the ^{nat}Kr concentration during SR2, so we just extrapolate from the end of SR1.

Materials A background component also arises from γ emissions from radioimpurities in detector materials that induce Compton-scattered electrons; however, with typical mean free paths of O(mm) these γs rarely traverse the $O(10)$ cm needed to reach the strict fiducial volume, scatter, and leave undetected again. This component is thus quite suppressed. The rate is constrained by radioassay measurements [27] and predicted by simulations [25] to be 2.7 ± 0.3 events/(t·y·keV). Based on the results of Geant4 simulations, the materials background is modeled by a fixed, flat component that does not change with time.

^{83m}Kr During SR1, background events from ^{83m}Kr were observed due to a trace amount of its parent nucleus ^{83}Rb in the xenon recirculation system. ^{83m}Kr was purposefully injected many times throughout the life of XENON1T for calibration purposes, where in normal conditions this calibration source decays away with half-life of 1.8 h. However, if the parent ^{83}Rb were injected, which we presume happened (accidentally) due to a momentary malfunction of the source valve, the time scale would increase due to the 86.2-day half-life of ^{83}Rb, which is what we observe.

^{83m}Kr decays via a two-step internal conversion (IC) process, resulting in the emission of two monoergetic electrons that produces ERs. The first decay produces a 32.1-keV IC electron, followed shortly thereafter ($T_{1/2} \sim 154$ ns) by a second, 9.4-keV electron [126]. Being a two-step decay, many of these events are removed by the multi-site selections mentioned in Sect. 2.3; however, due to the short half-life of the second step, these decays are also often unresolved in time and hence contribute as a mono-energetic peak at 41.5 keV. Using ^{83m}Kr calibration data, we confirmed that the 32.1- and 9.4-keV decays are removed by the event selection. Thus, the expected background contribution from this source is a single monoergetic peak at 41.5 keV, with its model shown in Fig. 3.3 (left). Upon observing a large 41.5-keV peak in our background spectrum—which did not decay away within ~ hours with the source valve closed—we confirmed that the temporal behavior was consistent with a ^{83}Rb contamination. In Fig. 3.3 (right), we show the evolution of the event rate in the energy range of 33–50 keV. This was then fitted with an exponential-plus-constant function, giving a $T_{1/2} = 94.6 \pm 8.6$ days and constant background rate of $b = 70.9 \pm 25.1$ events/(t · y · keV), consistent with the ^{83}Rb half-life and our expected background from ^{214}Pb. Also extracted from the fit is an initial ^{83m}Kr rate

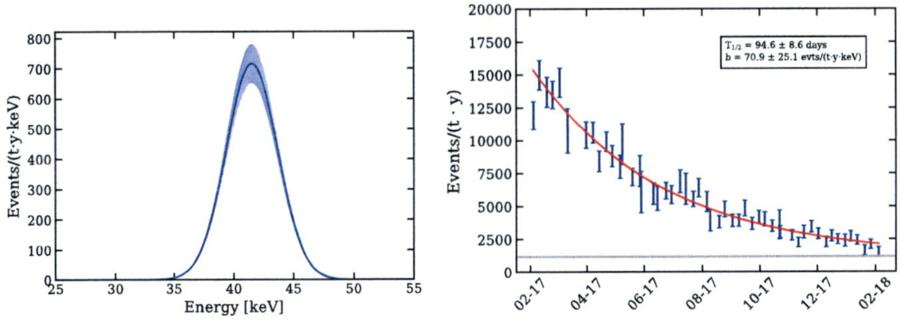

Fig. 3.3 Background model for ^{83m}Kr in SR1. Left: energy spectrum, a Gaussian centered at 41.5 keV and resolution given by the empirical model in Eq. (2.7). Right: temporal evolution model based on an exponential-plus-background fit to the rate in 33–50 keV energy range as a function of time. The temporal model is used for the constraint on the overall event rate shown in the left-hand plot

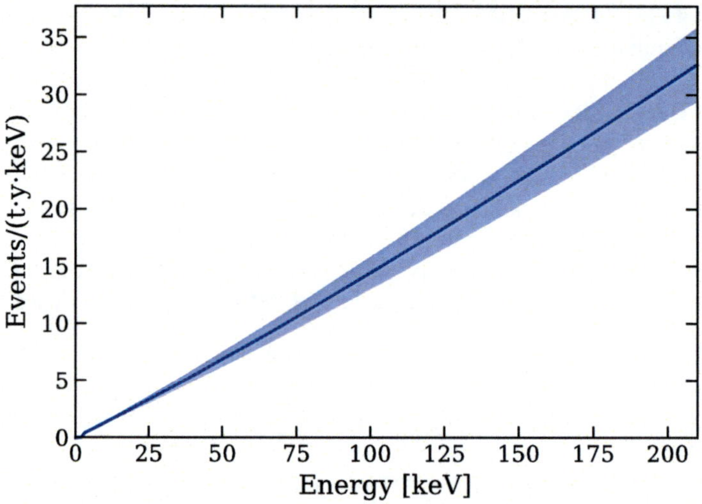

Fig. 3.4 Background model for ^{136}Xe in SR1, which is constant in time

of 15400 events/(t·y) at 0400 UTC, February 4, 2017. The fit gives a $\chi^2/\mathrm{ndf} = 1.1$, suggesting the fit function describes well the data.

The spectrum in Fig. 3.3 (left) and temporal evolution (right) summarize the ^{83m}Kr model for the SR1 analysis, where uncertainties on the final overall rate were extracted using toy-MC methods and used as a Gaussian constraint. For the SR1 run selection, this uncertainty is 9.0%.

^{136}Xe ^{136}Xe is a long-lived ($T_{1/2} \sim 10^{21}$ y) $2\nu\beta\beta$ emitter intrinsic to xenon, and thus irreducible (apart from isotopic depletion, which was not performed for XENON1T). The two βs emitted by this decay scatter ∼ immediately, leading to a single S1+S2 event given by the sum of the two β energies. The energy spectrum in the ROI can be found in Fig. 3.4. It was constrained in the fit according to the predicted rate and associated uncertainties on (1) a ^{136}Xe isotopic abundance of $(8.49 \pm 0.04_{\mathrm{stat}} \pm 0.13_{\mathrm{sys}})\%$ as measured by a residual gas analyzer [93], (2) the reported half-life [15], and (3) the spectral shape [120, 121]. We note that the spectrum in [120, 121] uses the so-called higher-state dominated (HSD) model, as opposed to the single-state dominated (SSD), where the lowest energy level dominates the decay. Small spectral differences at low energy are expected between the two models, as seen in ^{82}Se [49], but at the time no calculation has been published for the SSD model for ^{136}Xe.[1] Since ^{136}Xe is sub-dominant at low energies anyways, we account for this additional uncertainty by increasing the rate

[1] Though we were kindly given a calculated spectrum from Jenni Kotila, Yale University. We revisit this at the end of Chap. 4.

uncertainty to a total of 10%. If we only included the contributions from (1) and (2), the total uncertainty would be 3.2%.

Being an intrinsic background, ^{136}Xe has no time dependence. It is thus modeled as constant-in-time.

^{131m}Xe We include three background components that were produced after neutron calibration, the first of which being ^{131m}Xe (IC). This source induces a monoenergetic ER background at 163.9 keV [116], is activated by neutron capture on the xenon, and decays with a half-life of 11.8 days. A time-dependence study was done similarly to the ^{83m}Kr case above, fitting the rate evolution with an exponential-plus-background model, with the added complexity of multiple discontinuous injections at known dates of neutron calibration.

This fit gives a measured ^{131m}Xe half-life of $T_{1/2} = 11.8 \pm 0.1$ days, consistent with the literature value, and a $\chi^2/\mathrm{ndf} = 72.6/100$. The resulting model in both energy and time space is shown in Fig. 3.5, where the uncertainty is extracted using toy-MC methods, similar to the ^{83m}Kr case above.

^{133}Xe ^{133}Xe, another activated background, decays via a β to an excited state of ^{133}Cs, which in turn de-excites via an 81 keV prompt γ [117] within O(ns). These two decays are unresolved, resulting in a continuous ER spectrum starting at ~75 keV, after accounting for the energy resolution. The time dependence was fitted and modeled very similarly to ^{131m}Xe, with the resulting model summarized in Fig. 3.6. In time regions near neutron calibration, this component dominates the high-energy range of this analysis.

^{125}I The third activated isotope ^{125}I, a daughter of ^{125}Xe, decays via EC with a 100% branching ratio to an excited state of ^{125}Te, which de-excites via a 35.5 keV γ. On account of the captured electron from a K-shell, L-shell, or M-shell with

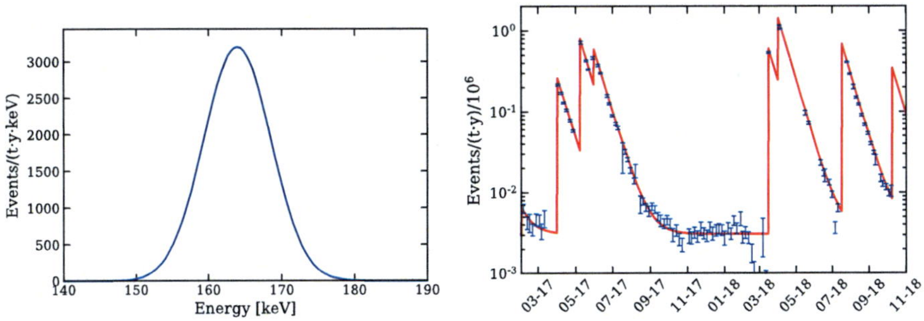

Fig. 3.5 Background model for ^{131m}Xe in SR1. Left: energy spectrum, a Gaussian centered at 163.9 keV and resolution given by the empirical model in Eq. (2.7). Right: temporal evolution model based on an exponential-plus-background fit to the rate within 4 standard deviations of the expected peak mean. The temporal model is used for the constraint on the overall event rate shown in the left-hand plot

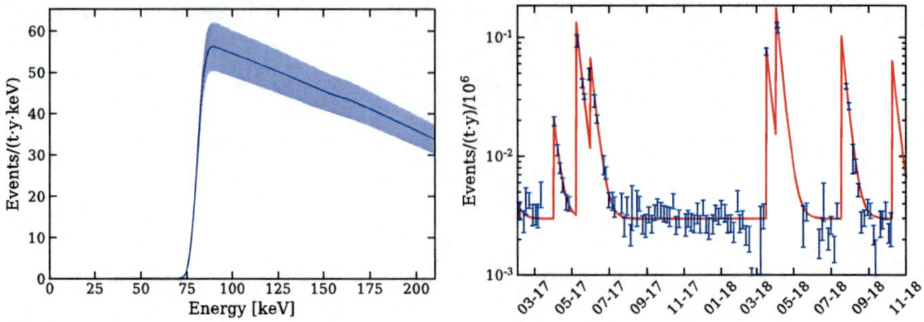

Fig. 3.6 Background model for ^{133}Xe in SR1. Left: energy spectrum. Right: temporal evolution model based on an exponential-plus-background fit. The temporal model is used for the constraint on the overall event rate shown in the left-hand plot

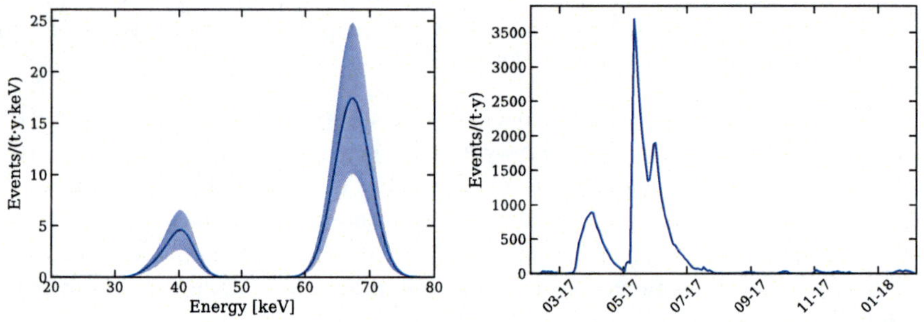

Fig. 3.7 Background model for ^{125}I in SR1. Left: energy spectrum, including three monoenergetic peaks at 67.3, 40.4, and 36.5 keV with fixed branching ratios. Right: temporal evolution model found by interpolating the background model described in detail in [34]. The temporal model is used for the constraint on the overall event rate shown in the left-hand plot

decreasing probability, the ^{125}Te γ is accompanied by an X-ray of 31.8, 4.9, or 1.0 keV, respectively. These two emissions are detected together, resulting in monoenergetic peaks at 67.3, 40.4, and 36.5 keV, respectively [54]. All three peaks of ^{125}I EC are included in the background model with the fixed branching ratios from [54]. The ^{125}I contribution was constrained using a model based on the time evolution of ^{125}Xe throughout SR1, as detailed in [34]. The ^{125}I model is summarized in Fig. 3.7. The ^{125}Xe energy spectrum is outside the ROI for this analysis and thus not relevant here.

^{124}Xe The first observation of two-neutrino double electron capture (2νECEC, or DEC) of ^{124}Xe was recently reported using the same SR1 dataset (but different event selection) as used in this analysis [34]. We consider this process as a background here. In 2νECEC decay two inner-shell electrons are simultaneously captured when two protons in the nucleus convert to neutrons, producing two neutrinos in the process that go undetected. In the case of ^{124}Xe, the resulting nucleus is the stable ^{124}Te. The electron shell vacancies induce a cascade of X-rays that are detected

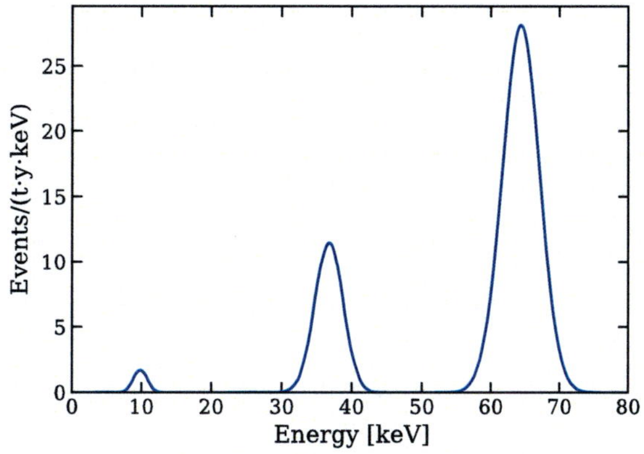

Fig. 3.8 Background model for 2νECEC of ^{124}Xe in SR1 including three monoenergetic peaks at 64.3, 36.7, and 9.8 keV with fixed branching ratios. The overall normalization of the three peaks was unconstrained in the fit, but the predicted values are given by the half-life calculation in [161]. Being an intrinsic background to xenon, this component is constant in time

together, resulting in a single ER with total energy deposited of (close to) the sum of the two electron binding energies.[2]

In [34] we considered the dominant branching ratio of 2νECEC, the capture of two K-shell electrons inducing a peak at 64.3 keV. It is also possible to capture a K-shell and L-shell electron (36.7 keV) or two L-shell electrons (9.8 keV) with decreasing probability, as calculated in [89]. For this analysis, the event selection and consideration of time dependence allow us to include all three peaks in the background model, as shown in Fig. 3.8. The predicted rates of the peaks are taken from an updated half-life [161] with fixed branching ratios from [89]; the overall rate was not constrained in the fit since the half-life was derived from the same dataset.

Solar Neutrinos Elastic scattering of solar neutrinos off electrons is expected to contribute subdominantly over the entire ROI. The expected energy spectrum was obtained using the standard neutrino flux in the Large Mixing Angle Mikheyev–Smirnov–Wolfenstein (LMA-MSW) model and cross section given by the SM [50, 112], with a step-function approximation applied to account for the effect of bound electrons at low energy. Based on rate calculations of neutrino-electron scattering in xenon as given in [69], a 3% uncertainty was assigned and used to constrain the solar neutrino rate in the fit. The model is shown in Fig. 3.9.

Due to changing distance between the Earth and the Sun, the neutrino flux in the detector is expected to modulate by a subtle ~7% peak-to-peak. However, since this background component contributes very little this modulation is ignored.

Other Backgrounds Due to the sub-dominant nature in our event selection, we ignore all NR backgrounds, accidental coincidence (AC) backgrounds, and so-

[2] The "double" nature of this decay increases the energy slightly. For the KK-capture of ^{124}Xe the energy is 64.3 keV, while the sum of the two ^{124}Te K-shells would be 31.8+31.8=63.6 keV [132].

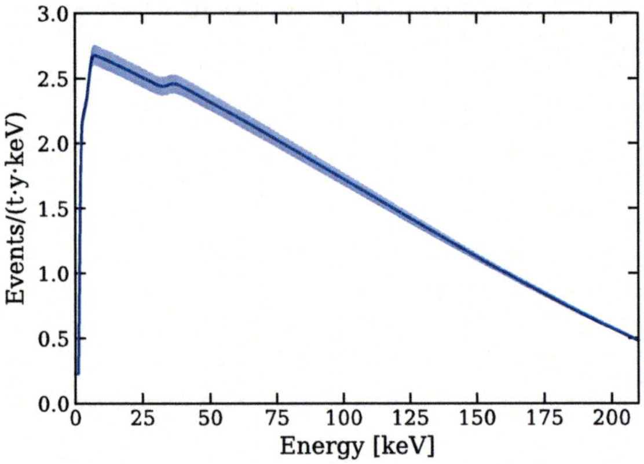

Fig. 3.9 Background model for elastic scattering of solar neutrinos on electrons. The step-function approximation used to account for electron binding energies can be seen at ∼30 and ∼5 keV. This component is expected to modulate in time according to the changing distance between the Sun and the Earth

called "surface" backgrounds due to event mis-reconstruction near the edges of the detector. More details on these backgrounds can be found in [36]. As described later, we will also consider two additional potential backgrounds: ^{37}Ar and ^{3}H (tritium). We do not have independent confirmation nor strong evidence of the presence of these backgrounds, but we considered them after comparing the SR1 data to the background model and so treat them separately.

Partitioning of the SR1 Dataset As mentioned briefly above, we partition SR1 into two sub-datasets. This is done on account of the activated background components, which display a strong time dependence during SR1. We define $\mathrm{SR1}_a$ to include SR1 events within 50 days of a neutron calibration campaign, and $\mathrm{SR1}_b$ to contain the rest. Accordingly, $\mathrm{SR1}_a$ is expected to display a much larger background rate from the activated component components, while $\mathrm{SR1}_b$, the longer of the two partitions, is expected to be much cleaner. For each background component, we use its time dependence model (and uncertainty) to predict its overall rate in the partition in question. Concerning backgrounds that are constant in time, their rates are fixed to be the same in both partitions, as described below in Sect. 3.3. For illustration, the predicted background spectra in the two partitions are summarized in Fig. 3.10, where the uncertainties on each component have been omitted for simplicity.

Summary of Background Model We denote the full, partitioned background model described above as B_0. This model includes a rate parameter for each component, except for the fixed materials background, that scales the rate normalization, constrained when applicable by Gaussian penalty terms. The ^{214}Pb and ^{124}Xe components were unconstrained. Additionally, shape parameters (colloquially referred to as *peak shapers*) for the dominant monoenergetic backgrounds ($^{83\mathrm{m}}$Kr, $^{131\mathrm{m}}$Xe, KK-capture of ^{124}Xe) were included that let the respective peak mean float within 1% of the expected value, a number chosen due to uncertainty on g_1 and g_2. The sub-dominant peaks did not have peak shapers to increase computation speed. Lastly, an additional shape parameter known as the *efficiency shaper* was used to scale the

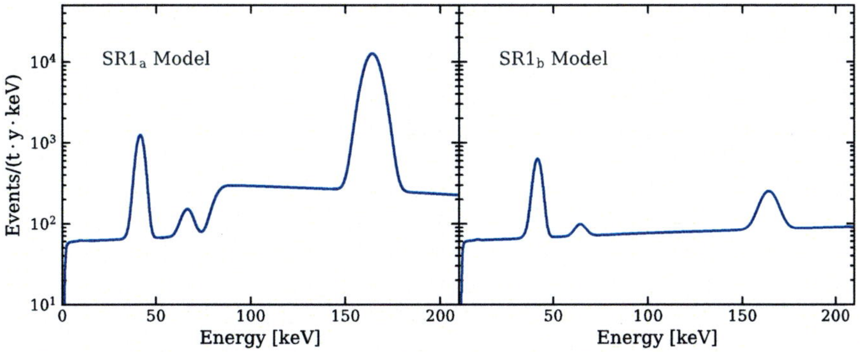

Fig. 3.10 Summary of the predicted background spectra in the partitioned SR1 dataset. Due to neutron activation, the background rate for much of the energy region is higher in $SR1_a$, whereas $SR1_b$ is relatively much cleaner. These two datasets will be fit simultaneously, with constant-in-time background shared between the two

efficiency curve up and down within the curve in Fig. 2.22, with a Gaussian prior. An efficiency shaper of $\pm$ 1 would correspond to the $\pm$ 1-σ bounds in Fig. 2.22. This model thus includes 9 rate parameters and 4 shape parameters and was used to fit the SR1 data in (1, 210) keV by maximizing the likelihood constructed in Sect. 3.3. In the next section, we consider the signal modeling.

3.2 Signal Modeling

In Chap. 1 we introduced the new-physics signals that we consider in this analysis: solar axions, an enhanced neutrino magnetic moment, and bosonic dark matter. Here we briefly describe how those signals would appear in XENON1T after including detector resolution and efficiency, just as for the background models in the previous section.

Solar Axions As discussed in Chap. 1, solar axions are produced through multiple interactions in the interior of the Sun, resulting in multiple components to the solar axion flux, three of which are considered here. The ABC flux scales with the square of the axion-electron coupling g_{ae}, the Primakoff flux with the square of the axion-photon coupling $g_{a\gamma}$, and the ^{57}Fe component with the square of an effective axion-nucleon coupling g_{an}^{eff}. After convolving these flux components with the axioelectric cross-section (Eq. 1.17), the event rates of all three components pick up an additional factor of ${g_{ae}}^2$, as summarized in Fig. 3.11.

The axion couplings to matter are in general model-dependent. For QCD axion models, these couplings are related to one another by the connection between the axion mass and the axion decay constant, as described in Sect. 1.2.1. For

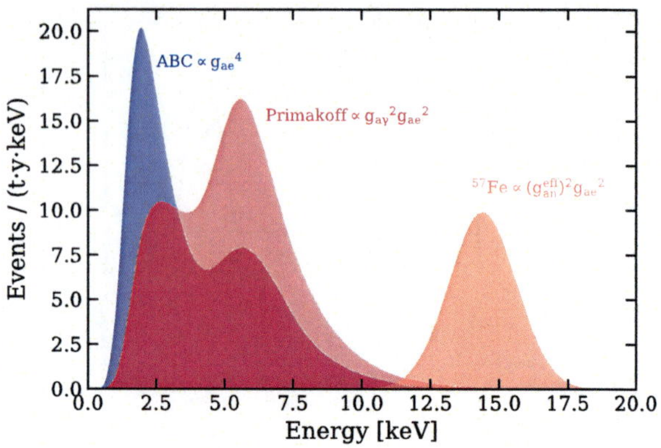

Fig. 3.11 Signal model for solar axions after accounting for detector efficiency and resolution. In this plot, the arbitrarily chosen coupling values are $g_{ae} = 3 \times 10^{-12}$, $g_{a\gamma} = 2 \times 10^{-10}$ GeV^{-1}, $g_{an}^{\text{eff}} = 10^{-6}$

example, the three couplings g_{ae}, $g_{a\gamma}$ and g_{an}^{eff} are completely determined by a single parameter (the axion mass m_a) in a hadronic axion model, or by two parameters (m_a and an additional term β_{DFSZ}) in a DFSZ model. However, axion-like particles (ALPs) in general do not have these model-dependent relationships; for these models the couplings to matter are taken as free parameters. Thus, in order to keep our analysis as model-independent as possible, we treat the normalizations of the three solar axion flux components as three, independent components of a single signal model.

The final signal model for solar axions, after convolving the energy spectra from Sect. 1.2.1 with our efficiency and resolution, is shown in Fig. 3.11. Again we stress that the three components are treated independently in the analysis, and are able to scale up/down freely with respect to each other in the fit. This three-component model is what we will search for on top of our background model B_0 in our solar axion search.

Neutrino Magnetic Moment and Bosonic Dark Matter The solar axion signal model is unique in this analysis due to the presence of three parameters. For the other two signal models, neutrino magnetic moment and bosonic dark matter, we simply convolve the true energy spectra from Sects. 1.2.2–1.2.3, where the neutrino magnetic moment spectrum is given in Fig. 1.7 and bosonic dark matter is a monoenergetic peak at the rest-mass of the particle.

For the neutrino magnetic moment, we are looking for an enhancement of the neutrino-electron scatter rate, especially at low energies, over the SM expectation from solar neutrinos. As we are already modeling the SM spectrum in our background model, our signal model for the neutrino magnetic moment only includes the electromagnetic enhancement. This is summarized in Fig. 3.12.

We omit a figure for the bosonic dark matter signal model due to its trivial nature as a monoenergetic peak with mean given by the mass of the particle and width given by detector resolution. It is noteworthy to mention that for this signal the spectrum depends on a nuisance parameter not included in B_0 (the particle mass), and so

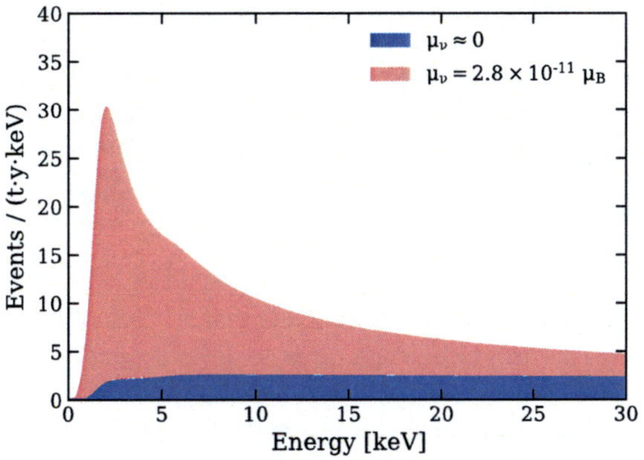

Fig. 3.12 Signal model for neutrino magnetic moment. As this signal would be an enhancement to the already-modeled solar neutrino background from the SM (purple), our signal model only includes the enhancement term (pink). This plot uses a value of $\mu_\nu = 2.8 \times 10^{-11}\ \mu_B$, the current world-leading upper limit from Borexino [7]

for this signal we need to account for the look-elsewhere effect in the statistical treatment. This is described in more detail in the next section.

3.3 Statistical Framework

The background and signal models from the previous two sections are the major inputs to the statistical inference, for which we use an unbinned profile likelihood analysis. The likelihood is constructed as

$$
\begin{aligned}
\mathcal{L}(\mu_s, \boldsymbol{\mu_b}, \boldsymbol{\theta}) &= \text{Poiss}(N|\mu_{tot}) \\
&\times \prod_i^N \left(\sum_j \frac{\mu_{b_j}}{\mu_{\text{tot}}} f_{b_j}(E_i, \boldsymbol{\theta}) + \frac{\mu_s}{\mu_{\text{tot}}} f_s(E_i, \boldsymbol{\theta}) \right) \\
&\times \prod_m C_{\mu_m}(\mu_{b_m}) \times \prod_n C_{\theta_n}(\theta_n), \qquad (3.1) \\
\mu_{\text{tot}} &\equiv \sum_j \mu_{b_j} + \mu_s,
\end{aligned}
$$

where

- μ_s is the parameter for the signal rate,
- $\boldsymbol{\mu_b}$ is a 9-dimensional vector representing the rates of all background components,
- $\boldsymbol{\theta}$ is a 4-dimensional vector representing the shape parameters; i.e., the efficiency and uncertainty on the peak locations of Gaussian backgrounds,
- index i runs over all N observed events, with E_i corresponding to the reconstructed energy of the ith event,

- index $j \in \{1, 2, 3, \ldots, 10\}$ runs over the background components, with f_{b_j} the reconstructed energy PDF for the jth background as given in Sect. 3.1,
- f_s is the reconstructed energy PDF for the signal,
- C_{μ_m} are Gaussian constraints on the applicable $m \in \{1, 2, 3, \ldots, 8\}$ backgrounds; i.e. all background except ^{214}Pb and ^{124}Xe, and
- C_{θ_n} are constraints on the $n \in \{1, 2, 3, 4\}$ shape parameters, Gaussian for the case of the efficiency and uniform between $\pm$ 1% for the peak shapers.

The main parameter we are interested in is μ_s, the strength of the signal rate. The others are nuisance parameters that are fit in the process of constraining μ_s.

As discussed above, due to time-dependent backgrounds the SR1 data set is divided into two partitions: SR1_a consisting of events within 50 days following the end of neutron calibrations and SR1_b containing the rest, with effective live times of 55.8 and 171.2 days, respectively. Including this time information allows for better constraints on the time-independent backgrounds and improves sensitivity to bosonic dark matter, especially, as the time-dependent background from ^{133}Xe impacts a large fraction of its search region. The full likelihood is then given by

$$\mathcal{L} = \mathcal{L}_\text{a} \times \mathcal{L}_\text{b}, \tag{3.2}$$

where $\mathcal{L}_\text{a}$ and $\mathcal{L}_\text{b}$ are evaluated using Eq. (3.1) in each partition. Nuisance parameters that do not change with time, along with all of the signal parameters, are shared between the two partitions. The constant nuisance parameters are:

- the efficiency parameter, which is dominated by detection efficiency and does not change with time.
- The ^{214}Pb component, which was determined to have a constant rate in time using detailed studies of the α-decays of the ^{222}Rn and ^{218}Po as well as the coincidence signature of ^{214}Bi and ^{214}Po.
- The solar neutrino rate, which would vary by ∼3 % between the two partitions on account of Earth's orbit around the Sun. This is ignored due to the subdominant contribution from this source.
- The decay rates of the intrinsic xenon isotopes ^{136}Xe and ^{124}Xe, as well as the Compton continuum from materials.

The remaining parameters all display time dependencies that are modeled in the two partitions.

The primary test statistic used for the inference is the *profiled log-likelihood ratio*, defined as

$$q(\mu_s) = -2\ln\frac{\mathcal{L}(\mu_s, \hat{\hat{\boldsymbol{\mu}}}_b, \hat{\hat{\boldsymbol{\theta}}})}{\mathcal{L}(\hat{\mu}_s, \hat{\boldsymbol{\mu}}_b, \hat{\boldsymbol{\theta}})}, \tag{3.3}$$

where the denominator $\mathcal{L}(\hat{\mu}_s, \hat{\boldsymbol{\mu}}_b, \hat{\boldsymbol{\theta}})$ is (3.1) evaluated at the overall set of signal and nuisance parameters that maximizes $\mathcal{L}$, while the numerator $\mathcal{L}(\mu_s, \hat{\hat{\boldsymbol{\mu}}}_b, \hat{\hat{\boldsymbol{\theta}}})$ is the evaluation of (3.1) at the values of the nuisance parameters that maximize $\mathcal{L}$ *for some input value of* μ_s. The likelihood ratio $q(\mu_s)$ thus quantifies the level of (dis)agreement between the null hypothesis $H_0 : \mu_s = \mu_s$ and the alternate hypothesis $H_1 : \mu_s = \hat{\mu}_s$ (the best-fit μ_s).

In a search for new physical processes, we are primarily interested in the null hypothesis $H_0 : \mu_s = 0$ and thus denote this special likelihood ratio as $q_0 = q(0)$. Excluding this hypothesis at a high significance is taken as evidence of new phenomena. Note that the q_0 test statistic is itself a random variable; for a given experiment/observation it takes on a value q_0^{obs}. For a realized measurement, the significance of rejecting the null hypothesis $H_0 : \mu_s = 0$, referred to as the *discovery significance*, is given by the p-value of the realized q_0^{obs}, i.e. the probability of having $q_0 \geq q_0^{\text{obs}}$ for an observed likelihood ratio q_0^{obs}. Determining this probability requires knowledge of the underlying distribution of q_0. Under asymptotic assumptions [17] and one signal parameter, the distribution can often be approximated as a χ^2 of one degree of freedom; however, in our framework we constrain μ_s to be ≥ 0, so the asymptotic distribution is modified to be the sum of a half-χ^2 distribution and half-δ function [74]. The asymptotic p-value for $\mu_s = 0$ is given by [74]

$$p = 1 - \Phi(\sqrt{q_0}), \tag{3.4}$$

where Φ is the cumulative density function (CDF) of the standard normal distribution.

It is customary in the physics community to quote p-values according to the number of standard deviations of the standard normal distribution that correspond to a given p. That is, the significance Z (for Z-score) is given by

$$Z = \Phi^{-1}(1 - p), \tag{3.5}$$

where Φ^{-1} is the inverse CDF of the standard normal. Combining (3.4) and (3.5), the significance of rejecting the null hypothesis is given by

$$Z = \sqrt{q_0}. \tag{3.6}$$

In particle physics, a significance of $Z = 5$ (often written 5σ) and absence of a compelling alternate explanation is typically required to claim discovery of a new phenomenon. In addition to determining the discovery significance, the test statistic q is also used for placing constraints on the value of μ_s. Concerning both the significance and the constraints on signal parameters, we treat each of the three signal models slightly differently.

As reviewed in [17], there are a set of conditions required for the asymptotic assumptions to be adequate.[3] In the context of this analysis, where the parameters of interest are the signal rates μ_s, these are:

1. *Sufficient number of events*. In this analysis this is satisfied, apart from the case of very large downward fluctuations in the background.
2. *Interior of allowed parameter space*. The values of the signal parameters μ_s and all nuisance parameters must not fall on the boundary of their allowed parameter space. Since the signal parameter is constrained to be ≥ 0, downward fluctuations of background (in which μ_s best-fit ~ 0) may lead to non-asymptoticity of the test statistic.
3. *Identifiable*. The energy spectra for different values of μ_s must be distinct from each other. This holds for the neutrino magnetic moment signal model, but not necessarily for the solar axion and bosonic dark matter models. The three-component solar axion model allows for similar *total* spectra with different parameters due to the low-energy degeneracy of the ABC and Primakoff components, while the dependence on bosonic dark matter mass introduces the look-elsewhere effect, where the same μ_s gives very different spectra for different mass values.
4. *Nested*. The null hypothesis is a limiting case of the alternate. This is satisfied, with the respective tests being $H_0: \mu_s = 0$ vs $H_1: \mu_s \geq 0$.
5. *Correct*. The true model is contained within one of the hypotheses tested. This is satisfied, as the signal is either present or it is not.

The neutrino magnetic moment signal model satisfies all conditions listed above, but the solar axion and bosonic dark matter cases do not. We accordingly handle the statistical inference for each signal model differently.

The neutrino magnetic moment signal model, with a single parameter that scales the rate of a well-defined spectrum up and down, is the simplest case. We assume asymptotic behavior of the test statistic, with significance given by Eq. (3.6). For the constraints on the magnetic moment μ_ν a modified Feldman-Cousins method described in [128] was adopted in order to derive 90% C.L. bounds with the right coverage. If the significance exceeds 3σ we will report a two-sided confidence interval on the magnetic moment μ_ν; otherwise, we will report a one-sided upper limit. This 3σ threshold was chosen before knowledge of the results in order to prevent reporting bias (a.k.a 'flip-flopping' [92]).

For the bosonic dark matter search, we account for the fact that we do not have a pre-defined value for the mass of the particle in question, which breaks condition 3 of the aymptotic assumptions described above. Since we search for many masses (every mass between 1 and 210 keV in 1-keV steps), we need to account for the fact that it is much more likely that we find a significant excess *somewhere* than if we only searched in one location. This is referred to as the "look-elsewhere effect" [103], which for this analysis was accounted for using toy-MC methods. In

[3] Originating from the famous theorem from Wilks [160].

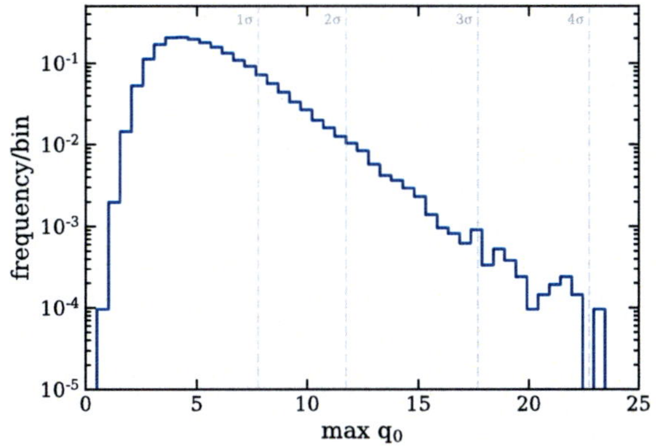

Fig. 3.13 Distribution of maximum q_0 values for the bosonic dark matter search, calculated using toyMC methods. This is used to account for the look elsewhere effect

the event of a peak-like excess, we would be most interested in the mass value that gave the largest significance and thus the largest value of q_0. To simulate this, many fake datasets were simulated under the background-only hypothesis, with each one being used to search for bosonic dark matter somewhere in the data, just as in the main analysis. For each simulation, after looping over masses and determining the q_0 for each one, the *maximum* value was tracked, with the distribution of $\max(q_0)$ from all the simulated datasets used as the reference distribution for determining discovery significance. The result is shown in Fig. 3.13. While a *local* significance can still be calculated using Eq. (3.6), the more relevant *global* significance is found by comparing a realized q_0^{obs} to the distribution in Fig. 3.13, giving a p-value that can be converted to a significance via Eq. (3.5).

Similarly to the neutrino magnetic moment search, for the bosonic dark matter analysis we use the approach in [128] to switch from one-sided to two-sided intervals depending on the significance of the signal. For this case, however, we use a different reporting threshold than 3σ to account for the look elsewhere effect described in the previous paragraph. We chose a value $q_0 = 16$ for the reporting threshold in bosonic dark matter, which from Fig. 3.13 is slightly under 3σ global significance.[4]

The statistical analysis of the solar axion is more complicated than the magnetic moment and bosonic dark matter cases. This model has three signal rate parameters: μ_{ABC}, μ_{Prim}, and μ_{Fe57}, which in turn scale with the axion couplings via g_{ae}, $g_{ae}g_{a\gamma}$, and $g_{ae}g_{an}^{\text{eff}}$, respectively. Due to the degeneracy at low energies between the ABC and Primakoff components (see Fig. 3.11), conditions 2 and 3 above may both be broken, implying that we cannot use asymptotic assumptions. Furthermore,

[4] Early in the analysis a threshold of $q_0 = 16$ was chosen as the reporting threshold because it corresponded to precisely 3σ global significance. Updates to the analysis changed the global significance distribution, but due to computational limitation the reporting threshold was not changed. It is anyway an arbitrary choice.

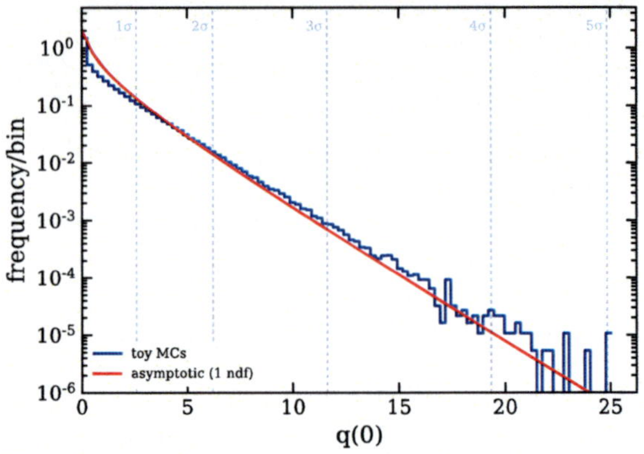

Fig. 3.14 Null distribution of discovery significance q_0 test statistic for the 3-component solar axion model (navy). The asymptotic distribution is shown in red to show poor agreement, and the dashed line show the conversion of the p-values to the conventional "sigmas" of the standard normal distribution

with the three signal rate parameters related to three axion couplings we would like to constrain, a generalized three-dimensional statistical analysis is warranted.

For a general solar axion model, the test statistic $q = q(\mu_{\text{ABC}}, \mu_{\text{Prim}}, \mu_{\text{Fe57}})$ is a three-dimensional function. As we cannot rely on asymptotic assumptions for the discovery significance, we employ toy MC techniques in a somewhat similar fashion to the bosonic dark matter case. Many background-only datasets were simulated, where for each one a likelihood ratio test between a B_0 + axion model and a B_0-only one was performed in order to build a distribution of $q_0 = q(0, 0, 0)$ values. This distribution, shown in Fig. 3.14, is used to determine the discovery p-value for the solar axion search. For illustration, the asymptotic distribution is also shown, with poor agreement validating the use of the toy MC approach instead of assuming asymptotics.

Due to the solar axion model's three-dimensional q, we wish to extend constraints on the signal rates to a 3-dimensional confidence *volume*. However, the modified approach for confidence intervals of [128] does not generalize to 3D. Therefore, for the solar axion search we instead use a standard profile likelihood construction, with no analogous reporting threshold of upper vs. central limits as there is for the magnetic moment and bosonic dark matter signals. This construction requires knowledge of the true distribution of q *for every point in the three dimensional space of* $\{\mu_{\text{ABC}}, \mu_{\text{Prim}}, \mu_{\text{Fe57}}\}$. Assuming this distribution is known, the 90% confidence volume is, by definition, the set of all rate parameters such that $q(\mu_{\text{ABC}}, \mu_{\text{Prim}}, \mu_{\text{Fe57}}) < q_{90}(\mu_{\text{ABC}}, \mu_{\text{Prim}}, \mu_{\text{Fe57}})$, where q_{90} represents the 90th percentile of the q distribution at that point in rate parameter space. We refer to q_{90} as the *critical value*, as for each point in the space of rate parameters, this value is the threshold by which to include/exclude that point in the confidence volume.

Since we cannot use asymptotic assumptions for the true distribution of q (and hence q_{90}), we again use toy MC techniques to determine it. For each point on a 3D grid of true $\mu_{\text{ABC}}, \mu_{\text{Prim}}, \mu_{\text{Fe57}}$, toy datasets were simulated with axion signal injected according to the rate parameters at that point. We denote the true rate parameters for a given point as $\hat{\mu}_s$. At each point, the true distribution of q was deter-

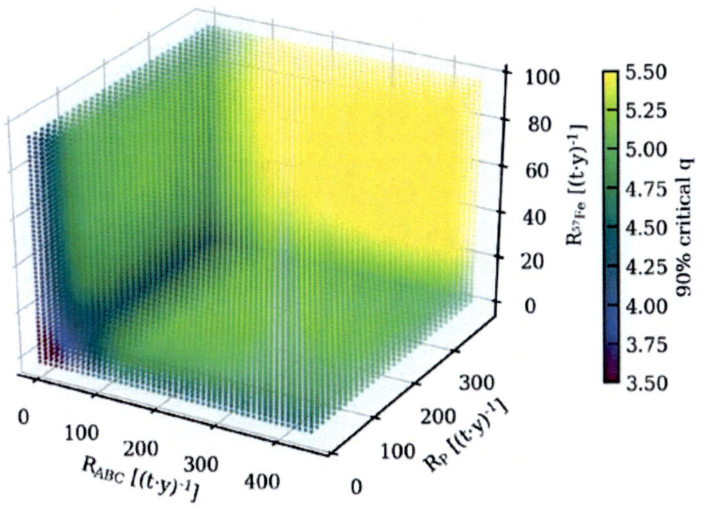

Fig. 3.15 Critical value threshold values for the solar axion search, as a function of ABC, Primakoff, and ^{57}Fe rate multipliers. For a given dataset, a 3D confidence volume is drawn out by the intersection of our test statistic q for that dataset with this 3D critical value function

mined by evaluating for each toy dataset the likelihood ratio $q(\hat{\mu}_{\text{ABC}}, \hat{\mu}_{\text{Prim}}, \hat{\mu}_{\text{Fe57}})$. The 90th percentile of this q distribution was then determined, giving the critical value q_{90} (again, just at that point). Repeating this procedure over the grid of true rate parameters defines a 3D grid of q_{90} values, which is then interpolated to define a 3D critical value function for a general set of $\mu_{\text{ABC}}, \mu_{\text{Prim}}, \mu_{\text{Fe57}}$. This interpolated function is shown in Fig. 3.15. By construction, the points for which $q(\mu_{\text{ABC}}, \mu_{\text{Prim}}, \mu_{\text{Fe57}}) < q_{90}(\mu_{\text{ABC}}, \mu_{\text{Prim}}, \mu_{\text{Fe57}})$ defines a three-dimensional 90% C.L. volume in the space of the three axion parameters. In Sect. 4.5 we report this volume evalulated with the SR1 data along with its two-dimensional projections, found by profiling over the third respective signal component.

In the next chapter, we apply this statistical framework to the real SR1 data.

Chapter 4
Results

Abstract In this chapter we present the results of our statistical inference applied to the XENON1T data selection discussed in the preceding chapters. We first describe the background-only fit to the data, summarizing the level of agreement and whether the different components are fitted as expected. A discrepancy between the background model and the data motivate further scrutiny and consideration of new potential background sources, specifically tritium. After this investigation, we then consider new physics potential explanations for the excess and extract constraints on the three signal models.

4.1 Background-Only Fit

The SR1 energy spectra for $\mathrm{SR1}_a$, $\mathrm{SR1}_b$, and their sum can be found in Fig. 4.1, along with the best-fit results under the background-only hypothesis. The summed best-fit values for the background components are summarized in Table 4.1, where the expected and fitted number of events are found by integrating over the 1–210 keV ROI the B_0 expectation and fit result, respectively, for each background component. All fit results for the different background components are consistent with our expectations, suggesting that we understand well the known sources of background.

From the B_0 fit we can extract a measured value of the ^{214}Pb concentration. Using the fitted number of events, total exposure, and branching ratio of 0.11 ± 0.01, we extract a total concentration of 11.1 ± 0.2 μBq/kg during SR1. This is within the upper/lower bounds of $(5.1 \pm 0.5, 12.6 \pm 0.8)$ μBq/kg found in Sect. 3.1.

We can also use the B_0 fit to estimate the total background rate. Integrating the best-fit from 1–30 keV, we find a total low-energy background rate of $76 \pm 2_{\mathrm{stat}}$ events/(t · y · keV), where the statistical uncertainty comes from integrating the spectra where each component's $\pm 1 - \sigma$ rate multiplier was used. This does not account for correlations between the rate multipliers, and thus slightly over-estimates the uncertainty. This is the lowest background rate ever achieved in this energy region, which as we will see allows XENON1T to have world-leading sensitivity to the electronic recoil signals presented here.

E. Shockley, *Study of Excess Electronic Recoil Events in XENON1T*,
Springer Theses, https://doi.org/10.1007/978-3-030-87752-1_4

Fig. 4.1 Background-only fit to the partitioned SR1 dataset. The two partitions in the lower two panels are fit simultaneously, with the summed data and fit results in the top panel. We observe a good fit to the data over most of the energy range

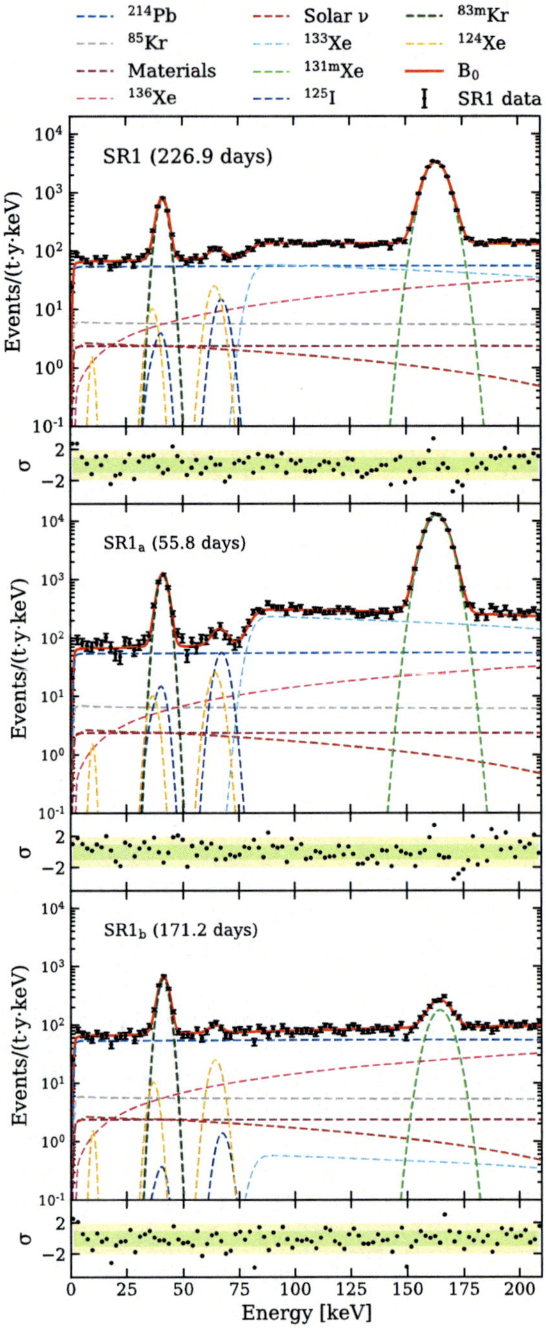

Table 4.1 Summary of components in the background model B_0 with expected and fitted number of events in the 0.65 tonne-year exposure of SR1. See text for details on the various components

Background components		
Component	Expected events	Fitted events
^{214}Pb	(3450± 340, 8530 ± 540)	7480 ± 160
^{85}Kr	890 ± 150	773 ± 80
Materials	323 (fixed)	323 (fixed)
^{136}Xe	2120 ± 210	2150 ± 20
Solar neutrino	220.7 ± 6.6	220.8 ± 4.7
^{133}Xe	3900 ± 410	4009 ± 85
^{131m}Xe	23760 ± 640	24270 ± 150
^{125}I (K)	79 ± 33	67 ± 12
^{125}I (L)	15.3 ± 6.5	13.1 ± 2.3
^{125}I (M)	3.4 ± 1.5	2.94 ± 0.50
^{83m}Kr	2500 ± 250	2671 ± 53
^{124}Xe (KK)	125 ± 50	113 ± 24
^{124}Xe (KL)	38 ± 15	34.0 ± 7.3
^{124}Xe (LL)	2.8 ± 1.1	2.56 ± 0.55

Goodness of Fit and Excess Events Since this is a maximum likelihood analysis, the goodness-of-fit is not as straightforward to evaluate as, e.g., a reduced-χ^2 analysis. Some texts suggest using a toy-MC approach of finding the distribution of likelihood values when simulating datasets under the best-fit parameters [73]; however, for our unbinned analysis this approach is not very sensitive to mismodeling on account of the likelihood values being dominated by the huge number of events in the localized ^{131m}Xe peak, essentially weighing the likelihood to describe a single energy region. We thus use a slightly different approach, still using toy-MC methods. Many toy SR1$_a$ and SR1$_b$ datasets were simulated and fit simultaneously, where the best-fits for the two datasets were then combined into a single fit for that dataset, similar to Fig. 4.1. This *total* best-fit was then compared to the spectrum of the (similarly combined) toy dataset, with 1-keV binning, using a standard χ^2 test statistic. The resulting distribution of χ^2 values from these simulations is used as the reference to which to compare the fit of the SR1 data, with its own χ^2 value determined in the same way, to determine a goodness-of-fit p-value. In this way, while the fit was performed using an unbinned profile log-likelihood method, the goodness-of-fit is evaluated using binned data with a simple χ^2 test. To test goodness-of-fit in different energy regions, we build the χ^2 distribution for a low-energy range (1–30 keV), mid-energy range (30–100 keV), and high energy range (100–210 keV), as well as the full ROI.

We show the goodness-of-fit p-values for the B_0 model and different energy ranges in Table 4.2. The mid-energy region from 30–100 keV, with a p-value of 0.32, is consistent with what we would expect from our background model. Conversely, both the high-energy (100–210 keV) and low-energy (1–30 keV) regions have p-values that suggest the data are not explained well by the model.

Table 4.2 Summary of goodness-of-fit p-values for the B_0-only fit and the B_0 + axion fit

Energy range	B_0 p-value	B_0 + axion p-value
(1, 30) keV	0.037	0.172
(30, 100) keV	0.32	0.31
(100, 210) keV	0.003	0.005
(1, 210) keV	0.001	0.007

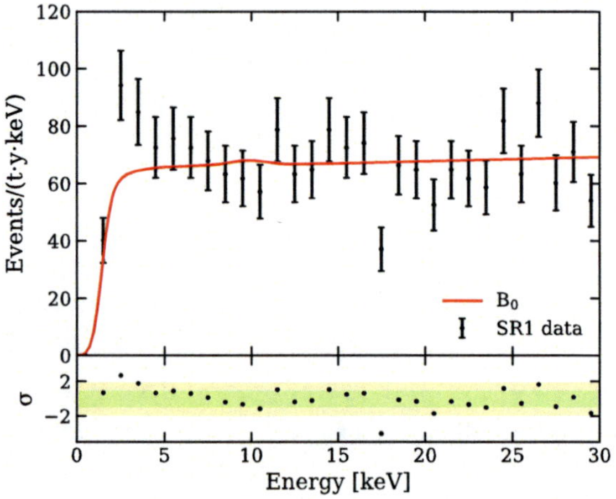

Fig. 4.2 Zoom of the background-only fit to the summed SR1 dataset, where an excess at low energies is observed

The high-energy region, with a p-value of 0.003, gives a particularly poor fit. This appears to be due to mismodeling of the dominant ^{131m}Xe background, where on account of the huge number of statistics even a minor systematic effect can lead to a large χ^2, due to the very small statistical uncertainty. Indeed, looking at the residuals in Fig. 4.1 in the region of the ^{131m}Xe peak at ~160 keV, there is a large discrepancy between the data and the model, while the rest of this energy range appears to agree quite well. The effect is most prominent during SR1$_a$ where the neutron activation was strongest. Since this mismodeling is localized to the region of a large monoenergetic background at high-energy, where we have little sensitivity to new physics anyway, we conclude that the poor p-value in the high-energy region is not a concern for the rest of the analysis.

The low-energy region, with a p-value of 0.037, hints that the background model might not describe well the data between 1–30 keV, albeit not nearly as extremely as the high energy region. Based on the residuals in Fig. 4.1, the poor fit appears to be in the ~1–10 keV region, where an excess of events is observed. It is notable that, being the ROI for the WIMP search, this low-energy region is the most studied in all of XENON1T. We show a zoom of this region in Fig. 4.2, where the data spectrum is seen to depart from the background model near ~7 keV and rise with decreasing energy, peak near 2–3 keV, and then subside to within 1σ of the background model near 1–2 keV. In contrast to the apparent ^{131m}Xe mismodeling, this excess is in a region where no spectral features are anticipated; the dominant ^{214}Pb background is by-and-large a flat spectrum. Between 1–7 keV, there are 285 events observed in the data compared to an expected 232 events from the background-only fit, implying an excess of $\gtrsim$50 events and a 3.3σ Poissonian fluctuation.

As an additional goodness-of-fit test we include a low-energy signal, the ABC solar axion,[1] in our model, now denoted as B_0 + axion. With this model, the p-value in the low-energy region improves to 0.172, a much better fit than the B_0-only model. This validates the intuition that the poor fit in the low-energy region is in fact due to the low-energy excess, and not, e.g., the single bin ~18 keV with strong deviation from the fit. It is worth noting that changing the binning of the data reduces the 18-keV fluctuation significantly, while the low-energy excess remains; however, as such *ex post facto* changes are not included in the goodness-of-fit simulations, we do not consider them for the SR1 dataset.[2] The p-values of the mid- and high-energy regions remain essentially the same as the B_0-only case, as expected when including only a low-energy source.

From the goodness-of-fit studies we conclude that (1) there are two regions that exhibit disagreement between the data and model, a low-energy region and high-energy region, which are completely unrelated to each other; (2) the high-energy discrepancy appears to be due to a mismodeling of the high-statistics ^{131m}Xe peak, which is not particularly concerning for the rest of the energy range; and (3) at low energy the data exhibit an excess at low energies relative to the background expectations.

The relatively poor fit in the low-energy region, with a $\gtrsim$ 3σ excess, is suggestive of a possible source of events not included in our background model, be it a systematic effect we missed, a new background, or perhaps new physics. In the following sections, we investigate the properties of this excess and its possible origins in more detail.

4.2 Investigating the Excess

Spatial Dependence Events within 1–7 keV appear to be uniformly distributed in the fiducial volume. In Fig. 4.3, we show the event distribution in z vs r^2 as well as x vs y, and differentiate between 2–4 keV and the rest of this reference region. Since it is not possible to disentangle on the event level the 'excess' from the background, it is difficult to determine the precise spatial distribution of the excess. We can, however, focus on the 2–4 keV region where the excess is most prominent, in which we find 70 events in the top half of the detector and 73 events in bottom half, which is consistent with a uniform hypothesis. These events are similarly uniform in the x

[1] This is just a proxy for "a low-energy component" to test how the goodness of fit changes and should be not be taken as evidence for a solar axion signal; for that we require the likelihood ratio tests, coming later in this chapter.

[2] We stress that these binning choices are only relevant for this goodness of fit study; for the likelihood ratio tests there is no binning involved whatsoever.

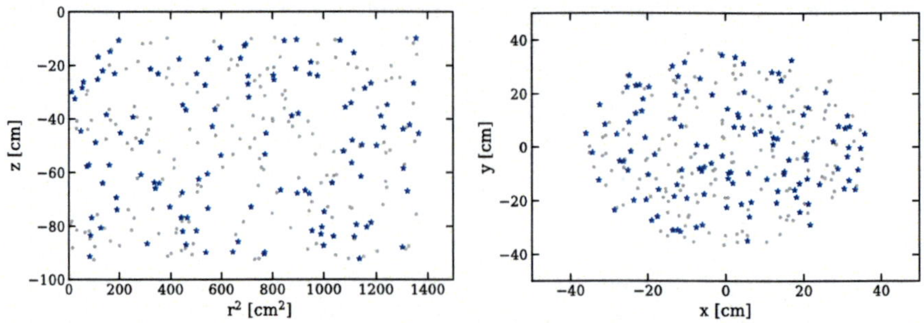

Fig. 4.3 Spatial dependence of SR1 events with energies between 1 and 7 keV. Events between 2 and 4 keV are highlighted in blue since this is most prominent region of the excess. The excess appears to be distributed uniformly in the detector

and y dimensions. Due to the strong excess-to-background ratio here, *if* the excess did have a strong spatial dependence, it bias these results.

Near our energy threshold, we would expect a position dependence due to non-uniform light collection efficiency in the TPC; events near the top of the detector produce smaller S1s than at the bottom, leading to relatively fewer events detected at the top. This would lead to a slightly non uniform distribution of events even in the case of a background perfectly unifrormly distributed in our fiducial volume. We investigated this with ^{37}Ar data by considering the S1 distribution of the 2.8-keV peak in different slices of z. Assuming a Gaussian S1 response for a given slice of z, we extract the efficiency by fitting the S1 distribution with a Gaussian convolved with our known efficiency in Fig. 2.20. Using the fit results of the mean and standard deviation, we estimate the expected efficiency loss due to depth by comparing the integral of the pure (i.e., no efficiency applied) Gaussian and the efficiency-corrected one. We show a few of the resulting data+fits for this study in Fig. 4.4, where overall good agreement is observed. The results of this study suggest a $\sim$15% decrease in efficieny between the top and bottom of the TPC, for a 2.8-keV line source. The effect is expected to be stronger with decreasing energy.

We compare the expected efficiency loss to the observed z distribution in Fig. 4.5, treating separately background events with 2–4 keV vs 4–7 keV. While the underlying background distribution is surely not a 2.8-keV line source, this efficiency should be a good estimate of the expected z-dependence in the energy region where the excess is most prominent. We conclude from this plot that the z distribution is consistent with uniform, though the uncertainty is quite large due to the low statistics.

Time Dependence The time dependence of the low-energy region was also investigated and determined to be inconclusive, as it is statisitically consistent with many hypotheses, including constant in time. We describe the time dependence in more detail in Sect. 4.9, after discussing different hypotheses for the excess.

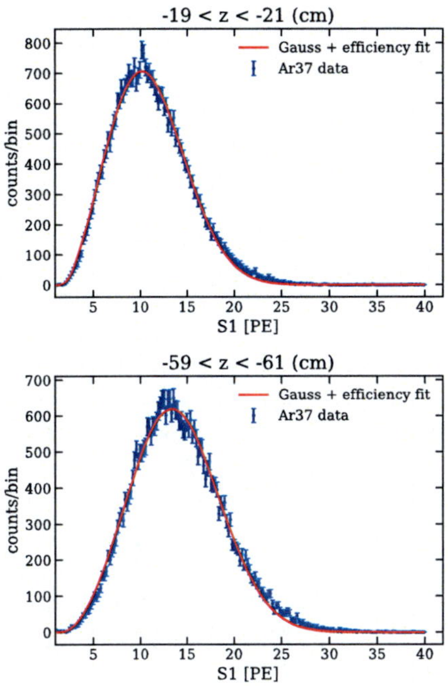

Fig. 4.4 Example fits of Gaussian + efficiency fits to the 2.8 keV peak in ^{37}Ar calibration data. The efficiency is fixed to the mean curve from Fig. 2.20. The fits agree with the data quite well, except for at low energy where the data display fewer events than expected in the low-energy tail. This behavior is not understood at this time, but hypothesized to be related to a selection cut

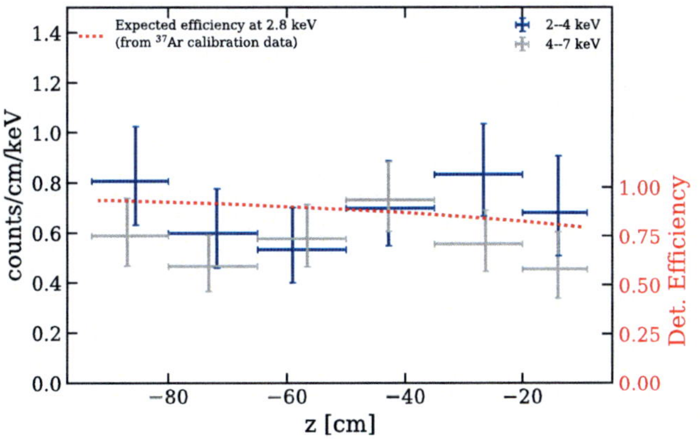

Fig. 4.5 Depth (z) distribution of low-energy background events, showing separately the 2–4 and 4–7 keV regions. The expected efficiency of a 2.8-keV line source from a study of ^{37}Ar calibration data is also included (red), to show roughly the spatial dependence expected at low energies. We conclude from this that the low-energy region is consistent with uniform in z

Systematic Effects Here we consider systematic effects in the analysis that could potentially lead to an apparent excess at low energies. The effects we consider are

- mismodeling of the energy threshold,
- mismodeling of energy reconstruction,
- event selection cuts, and
- systematic background sources.

We discuss these four effects in detail in the following paragraphs, concluding with a study using the exact same analysis framework to fit ^{220}Rn calibration data and thus test all four systematic effects.

We explored the energy threshold and detection efficiency in detail in Chap. 2, where the efficiency was validated using both ^{220}Rn and ^{37}Ar calibration data, as well as with the FAX simulation. The S1 efficiency is further validated by Fig. 4.4, where the S1 distributions of ^{37}Ar in thin z slices agree well with Gaussians convolved with the S1 efficiency evaluated in Chap. 2. Based on these already-described studies, we conclude that the detection efficiency is well understood.

The observed low-energy excess is peaked above ~2 keV, where the detection efficiency of XENON1T is >80% (see Fig. 2.22). While this is certainly *near* our threshold, it is high enough to not be *right at* our threshold where mismodeling of the efficiency could lead to an apparently large excess in the residuals. Further, from Fig. 4.2 it is clear that this excess is not just an artifact of residuals anyway; the rate is significantly higher ~2–7 keV than $\gtrsim$10 keV; this cannot be explained by a threshold mismodeling unless the efficiency were to *decrease* with energy above 2 keV. This is not compatible with the study of ^{220}Rn data in Sect. 2.2, nor with the more detailed study with the same data described below.

We also studied the behavior of the cuts we apply to remove poorly reconstructed events, etc. If the excess were an artifact from the event selection, it would almost certainly be due to events that should have been removed but were not. However, for every parameter space in which our cuts are defined, the low-energy region behaves as expected. There are no suspicious clusters of events in any parameter space observed, nor are there large numbers of events near the thresholds of a cut definition. As described below, applying the exact same cuts to ^{220}Rn calibration data leads to no such excess. We also looked at the waveform of every single event below 7 keV; every one looked like a normal ER event. Considering there are ~50 events more than expected, we believe we can rule out issues with event selection.

The energy reconstruction was investigated using ^{37}Ar data, our lowest energy calibration source. Fig. 4.6 shows the comparison of the ^{37}Ar calibration spectrum, using the g_1/g_2 values from SR2 (see Sect. 2.1.1), and how this analysis models a 2.8 keV monoenergetic peak. Some nongaussianity is observed in the calibration spectrum, but the mean of the spectrum is 2.828 ± 0.002 keV, which is within 1% of the literature value of 2.822 keV [52] and thus validates the energy reconstruction at low energies. We note that this study also highlights some mismodeling of the resolution at low energies, as the observed ^{37}Ar spectrum is narrower than what we would expect based on the energy resolution model in Fig. 2.8. As discussed in

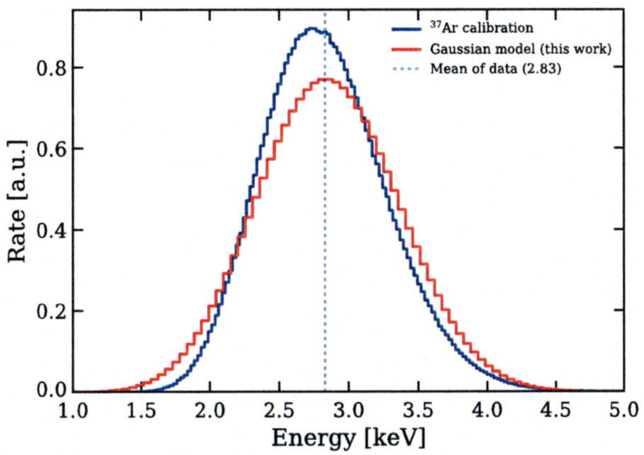

Fig. 4.6 Comparison of the ^{37}Ar spectrum from calibration data (navy) to how we model it in this analysis (red), both normalized to unity. Some nongaussianity is observed in the calibration spectrum, with the mean of the distribution offset from the peak. The mean is the more relevant value which, within 1% of the expected value, validates the energy reconstruction for this analysis

Sect. 2.1.2, such a small mismodeling of the resolution does not impact continuous energy spectra and thus would not affect the ^{214}Pb background expectation. It *might* have a small impact on the interpretation of monoenergetic peaks in this energy region, which we discuss in Sect. 4.7, but it would certainly not lead to a 50-event excess.

There are two main instrumental backgrounds in XENON1T: Accidental coincidences (AC) and the so-called "surface" background. The AC background arises due to the accidental pairing of unrelated S1 and S2 signals in the detector that happened to occur at the same time. Being real S1 and S2 signals—just uncorrelated ones—AC events can easily emulate real interactions in the xenon. Since they are random, these events are expected to be reconstructed as spatially uniform, and often as low-energy events due to their small S2s. However, the AC background was carefully studied for the WIMP analysis and constrained to have a rate of < 1 event/(t·y) based on the rates of 'lone' signals in the detector, S1s (S2s) that do not have a corresponding S2 (S1) [36]. The rate would be even further suppressed by the 500 PE S2 threshold used in this analysis. Lastly, from the same study of lone S1s/S2s, the distribution of AC in the space of S2-vs.-S1 is well-understood and not observed in the SR1 data, as shown in Fig. 4.7.

The surface background originates mainly from radon daughters plated out on the PTFE wall of the detector [36]. These events thus occur near the edge of the detector whereby, during the electron drift, field effects can degrade the electron cloud en route to the liquid-gas interface, leading to a poorly-reconstructed, smaller S2 signal. This background has a strong spatial dependence [36] and can be removed by fiducialization. Indeed this is one of the main reasons we use the 1.0 t volume for

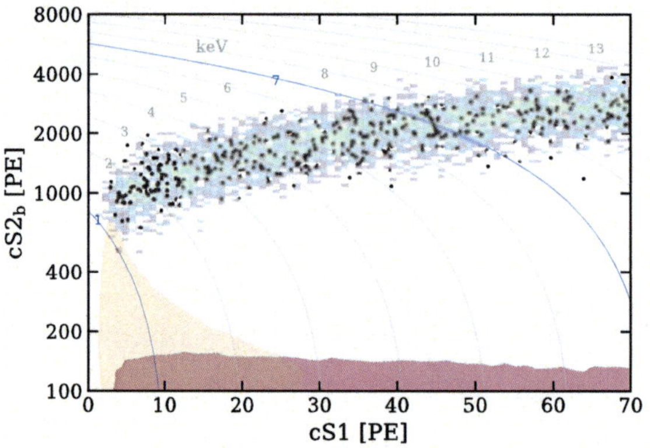

Fig. 4.7 Distribution of low energy events (black dots) in the ($cS1$, $cS2_b$) parameter space, along with the expected surface (purple) and AC (orange) backgrounds ($1\,\sigma$ band). ^{220}Rn calibration events are also shown (density map). All the distributions are within the one-tonne fiducial volume. Gray lines show isoenergy contours for electronic recoils, where 1 and 7 keV contours, the boundaries of the reference region, are highlighted in blue

this analysis, corresponding to a radial distance from the TPC surface of $\gtrsim$11 cm, as opposed to the larger 1.3 t used in others [31]. Similarly to the AC background, the strict S2 threshold further suppresses this background to a negligible level, and the anticipated distribution in S1-S2 space is not observed in Fig. 4.7. We thus rule out the surface background as a contributing factor to the excess.

As a final cross-check of all systematic effects, we use our entire analysis framework to fit ^{220}Rn calibration data. The β decay of ^{212}Pb, a daughter of ^{220}Rn, was used to calibrate the ER response of the detector, and thus allows us to validate our analysis with a high-statistics data set. Similarly to ^{214}Pb, the model for ^{212}Pb was calculated to account for atomic screening and exchange effects, as detailed in the Appendix of [40]. The same selection cuts were applied to the ^{220}Rn data as the SR1 data, including a 1 t fiducial volume. We then use the same statistics framework (Sect. 3.3), including the efficiency nuisance parameter, to fit the calibration data to the ^{212}Pb background model. This fit is shown in Fig. 4.8, where excellent agreement is observed (p-value = 0.50). Additionally, the S1 and S2 signals of the low-energy events in background data were found to be consistent with this ^{220}Rn data set, as shown in Fig. 4.7. In conclusion, if a systematic effect were to blame for the observed excess, it would have to be one that was not present for the ^{220}Rn calibration data, which is quite restrictive. There is no such effect that we can think of at this time.

Shape Uncertainty in Backgrounds Here we consider uncertainties in the calculated background spectra, particularly for the dominant ^{214}Pb background. A steep rise in the spectrum at low energies could potentially be caused by exchange effects; however, based on the calculation from [40], this component is predicted to be accurate to within 1% and therefore negligible with respect to the observed excess.

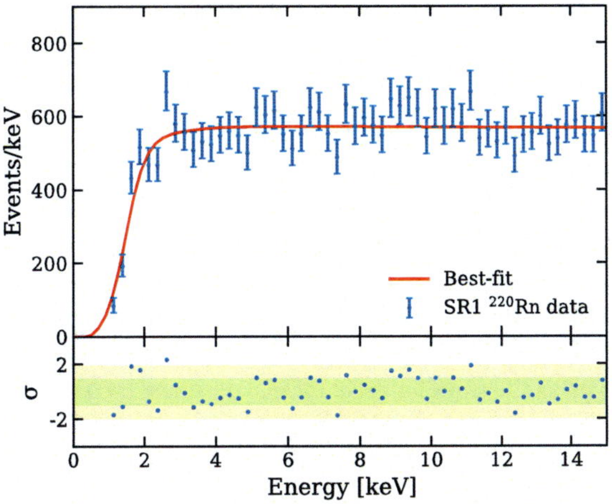

Fig. 4.8 Fit to ^{220}Rn *calibration* data with a theoretical β-decay model and the efficiency nuisance parameter, using the same unbinned profile likelihood framework described in Sect. 3.3. This fit suggests that the efficiency shown in Fig. 2.22 describes well the expected spectrum from ^{214}Pb, the dominant background at low energies

The remaining two factors, namely the endpoint energy and nuclear structure, tend to shift the entire β distribution, not cause steep changes over a range of ∼10 keV. Since we leave the ^{214}Pb component unconstrained, this uncertainty is largely accounted for; regardless, the combined uncertainty is +6% in the 1–10 keV region [40]. In comparison, a +50% uncertainty at 2–3 keV on the calculated ^{214}Pb spectrum, as constrained by the higher energy component, would be needed to make up the excess.

The sub-dominant backgrounds contributing at low energy are ^{85}Kr, ^{136}Xe, materials, and solar neutrinos. The uncertainty on ^{85}Kr is similar to the ^{214}Pb case above, but with a total contribution ∼1/10 of ^{214}Pb, would require a roughly +500% uncertainty to explain the excess. There is some uncertainty on the ^{136}Xe distribution due to the difference between HSD (used here) and the SSD spectrum, but, as we discuss in Sect. 4.10, this uncertainty goes in the opposite direction of what would be required to explain the excess. The SSD calculation (albeit an unpublished result) predicts a lower rate than the HSD model. The materials background from Compton scatters, even more suppressed than ^{85}Kr, is not expected to lead to such a large increase in a localized region; furthermore, this background would have a strong spatial dependence that we do not observe. Lastly, the solar neutrino background is well constrained, apart from the possibilities of new physics such as the neutrino magnetic moment. We consider this later in this chapter.

4.3 Potential Backgrounds

We also considered backgrounds that we did not include in B_0 but might in principle be present in trace amounts.

4.3.1 *Xenon X-Rays*

First, low-energy X-rays from ^{127}Xe EC, as seen in [11] and [95], are ruled out for a number of reasons. ^{127}Xe is produced from cosmogenic activation at sea level; given the short half-life of 36.4 days and the fact that the xenon gas was underground for O(years) before the operation of XENON1T, ^{127}Xe would have decayed to a negligible level. Indeed, high-energy γs that accompany these X-rays were not observed, and with their O(cm) mean free path in LXe they could not have left the O(m)-sized TPC undetected. For these reasons, we conclude that any ^{127}Xe cannot account for the excess.

4.3.2 *^{37}Ar*

Another potential background we considered is ^{37}Ar, which decays via EC to the ground state of ^{37}Cl, yielding a 2.82 keV peak with a 0.90 branching ratio [52]. Traces of ^{37}Ar were considered by the LUX collaboration to explain a possible excess rate at ~3 keV in their data [10]. While no definitive conclusion was ever published, the LUX collaboration hypothesized that an ingress of ^{37}Ar could originate from either (1) an initial amount in the xenon gas, or (2) an air leak during operation of the detector [12]. We follow suit and consider here these two possibilities for introduction of ^{37}Ar into XENON1T.

In order to account for the excess, the required ^{37}Ar rate is ~65 events/(t · y), based on a fit to the data with a 2.8 keV peak as shown in Fig. 4.9. ^{37}Ar has a half-life of $T_{1/2} = 35.0$ days [52] and a typical abundance in ^{nat}Ar of $\sim 10^{-20}$ mol/mol [146]. The initial measured ^{nat}Ar concentration in the xenon inventory was <5 ppm [110]. We therefore estimate that the *initial* rate before the start of XENON1T was

$$\begin{aligned} R_0^{\mathrm{Ar37}} &\lesssim \left(\frac{5\ ^{\mathrm{nat}}\mathrm{Ar\ atoms}}{10^6\ \mathrm{Xe\ atoms}}\right)\left(\frac{N_A\ \mathrm{Xe\ atoms}}{A}\right)\left(\frac{10^{-20}\ ^{37}\mathrm{Ar\ atoms}}{^{\mathrm{nat}}\mathrm{Ar\ atoms}}\right)\left(\frac{\ln 2}{T_{1/2}}\right) \\ &\lesssim 1.7\times 10^3\ \mathrm{events}/(\mathrm{t}\cdot\mathrm{y}), \end{aligned}$$

where N_A is Avogadro's number and A is the molar mass of xenon. The beginning of SR1 was ~400 days after the measurement of the xenon inventory, meaning that any ^{37}Ar remaining in the xenon would have decayed to, at most

$$\begin{aligned} R_{\mathrm{SR1}}^{\mathrm{Ar37}} &\simeq R_0^{\mathrm{Ar37}} e^{-(400\ \mathrm{days})\ln 2/T_{1/2}} \\ &\simeq 0.6\ \mathrm{events}/(\mathrm{t}\cdot\mathrm{y}) \end{aligned}$$

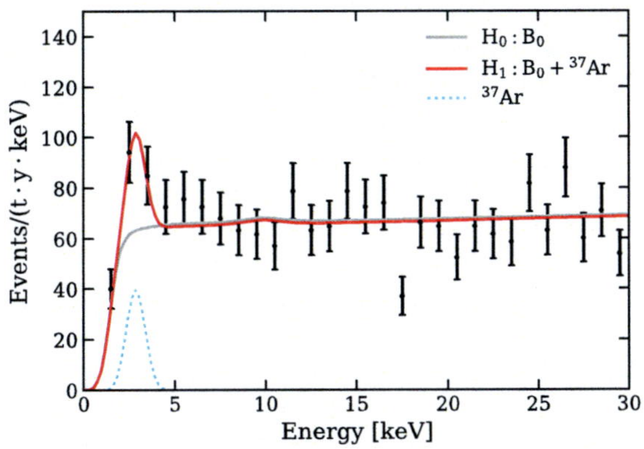

Fig. 4.9 Fit of the excess with an ^{37}Ar hypothesis. In order to explain the excess an ^{37}Ar rate of 65 events/(t·y) is required, based on the best-fit. As described in the text, such a rate from initial concentration prior to filling or from an air leak is excluded

While this is already a factor of ~100 too small to account for the excess, this initial argon concentration would also have been removed by online ^{85}Kr distillation. The removal of argon via online distillation was specifically tested in the final months of XENON1T (after all science data), during which ^{37}Ar was injected as a calibration source and then removed by distillation. The time constant of argon removal was measured to be ~2 days. With 90 days of online distillation occurring in the early days of XENON1T, any initial ^{37}Ar activity was further suppressed by a factor of $e^{-90/2} \sim 10^{-20}$, conclusively ruling out the presence of ^{37}Ar from its initial concentration in the xenon inventory.

Since there is a natural abundance of ^{37}Ar in the air, a constant air leak into the xenon handling system (if occurring after distillation) could be another source of ^{37}Ar. If such a leak were present, argon, a noble gas, argon would not be removed by the xenon purification system. This is also the case for krypton, which, as discussed in Sect. 3.1, is an important source of background and thus monitored carefully. From frequent measurements of the kypton concentration using rare gas mass spectrometry (RGMS) [124] and its natural abundance in the air, we can constrain the size of a hypothetical air leak. Over the course of SR1, the ^{nat}Kr concentration increased by an average 10.4×10^{-12} kg/d [131]. This increase is consistent with the expectation from outgassing of detector materials, but to be conservative we assume it is entirely due to an air leak. With a typical krypton concentration in air of ~1.1 ppmv [22], such an increase would correspond to an air leak of

$$\begin{aligned} L_{\text{air}} &= 10.4 \times 10^{-12}\,\text{kg/d} \left(\frac{365\,\text{d}}{\text{y}}\right) \left(\frac{\text{L}_{\text{Kr}}}{3.75 \times 10^{-3}\,\text{kg}}\right) \left(\frac{10^6\,\text{L}_{\text{air}}}{1.1\,\text{L}_{\text{Kr}}}\right) \\ &\simeq 0.9\,\text{L/y}, \end{aligned}$$

where 3.75×10^{-3} kg/L is the density of krypton.

The air inside the experimental hall at LNGS, supplied from outside of the laboratory and fully exchanged within 2.5 h, has an ^{37}Ar concentration of <3.2 mBq/m^3 (90% C.L.), determined from measurements taken in July 2020 following the methods in [142, 144]. We conservatively use for our estimation a value of 5 mBq/m^3 to account for possible seasonal variations [97, 143]. Assuming this activity and an air leak of 0.9 L/y from above, the number of ^{37}Ar atoms that would be introduced is given by

$$N^{\text{leak}}_{\text{Ar37}} = \left(\frac{5 \times 10^{-6}\ \text{Bq}}{\text{L}_{\text{air}}}\right)\left(\frac{35\ \text{d}}{\ln 2}\right)\left(\frac{86400\ \text{s}}{\text{d}}\right) \times (0.9\ \text{L}_{\text{air}}/y)$$
$$\simeq 20.1\ \text{atoms/y},$$

where the ^{37}Ar half-life was used to convert between activity and number of atoms.

However, the different viscosities of krypton and argon would affect their respective leak rates. An air leak of 0.9 L/y is within the laminar flow regime, in which the leak rate of a gas scales inversely with its viscosity η. With $\eta_{\text{Kr}} = 25.5 \times 10^{-6}$ Pa s and $\eta_{\text{Ar}} = 22.8 \times 10^{-6}$ Pa s, we accordingly should correct the number of atoms by a factor of $25.5/22.8 \simeq 1.1$.

Lastly, we need to consider the relative volatility of argon in liquid/gaseous xenon, as only decays in the liquid phase would contribute as a background. Again to be conservative, we take the relative volatility of argon $\alpha \sim 50$, though it is likely closer to 100 in XENON1T [42]. In XENON1T, there is 3.2t of xenon in the phase liquid and 21 kg in the gaseous phase, respectively. Including also the factor of 1.1 from the different viscosities, we therefore conclude with a final upper limit on the total rate ^{37}Ar decays in the liquid from an air leak of

$$R^{\text{leak}}_{\text{Ar37}} \simeq 1.1 \cdot (20.1\ \text{atoms/y})/(3.2\ \text{t} + 50 \times .021\ \text{t}) = 5.2\ \text{decays/(t} \cdot \text{y)},$$

assuming the leak rate has reached equilibrium with the ^{37}Ar decay.

To summarize, we measured with RGMS a very small increase in the concentration of krypton of $\lesssim$1 ppt/y over the course of SR1, which we conservatively assume is entirely due to an air leak. Assuming this conservative air leak, a conservative relative volatility of argon in liquid xenon, and equilibrium between the leak and the argon decay, we get an upper limit on the argon rate of 5 events/(t·y). This is more than an order of magnitude less than the 65 events/(t·y) required to make up the excess. This conservative constraint therefore excludes ^{37}Ar from a constant air leak as an explanation for the excess in SR1.

4.3.3 Tritium

We also considered an additional potential background that has never been observed before in LXe TPCs: the β emission of tritium[3] (^{3}H or T), which has a Q-value of 18.6 keV and a half-life of 12.3 years [125], and hence could the increasing

[3] Tritium in the form of tritiated methane has been used for calibration of LXe TPCs [9, 32, 75], including XENON100, but was not used as a calibration source in XENON1T. It is also worth

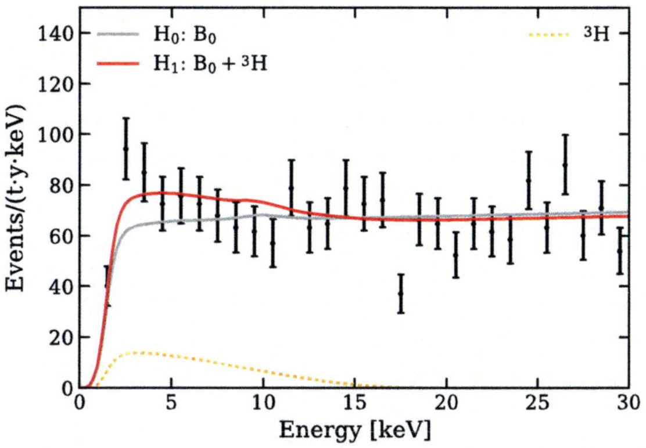

Fig. 4.10 Fit results for the tritium hypothesis (red) compared to the B_0-only hypothesis (gray)

spectrum at low energies. Tritium may be introduced from predominantly two sources: cosmogenic activation of xenon during above-ground[4] exposure [164] and emanation of tritiated water (HTO) and hydrogen (HT) from detector materials due to its cosmogenic and anthropogenic abundance.

In order to determine the hypothetical concentration of tritium required to account for the excess, we search for a ^{3}H 'signal' on top of the background model B_0, as shown in Fig. 4.10. The tritium rate suggested by the fit is 159±51 events/(t·y) (68% C.L.), corresponding to a ^{3}H/Xe concentration of

$$[T] = \left(\frac{159 \pm 51 \text{ events}}{\text{t} \cdot \text{y}}\right)\left(\frac{12.3 \text{ y}}{\ln 2}\right)\left(\frac{131.3 \times 10^{-6} \text{ t/mol}}{N_A}\right)$$
$$= (6.2 \pm 2.0) \times 10^{-25} \text{ mol/mol},$$

where the molar mass of xenon and the tritium half-life were used. This is $\lesssim 3$ tritium atoms for each kg of liquid xenon.

A tritium background component from cosmogenic activation of target materials has been observed in several dark matter experiments at rates compatible with predictions [18], although it has never before been detected in xenon. At sea level, the tritium activation rate in xenon is expected to be 31.6 $(\text{kg} \cdot \text{d})^{-1}$ [164] . From exposure to cosmic rays during above-ground storage of xenon, we estimate a conservative upper limit on the *initial* ^{3}H/Xe concentration, meaning the concentration when the xenon bottles were moved underground, of

$$[T]_0 < \left(\frac{31.6}{\text{kg d}}\right)\left(\frac{12.3 \text{ y}}{\ln 2}\right)\left(\frac{365 \text{ d}}{\text{y}}\right)\left(\frac{0.1313 \text{ kg/mol}}{N_A}\right)$$
$$= 4.46 \times 10^{-20} \text{ mol/mol}.$$

mentioning that neither the xenon used in XENON100 nor the materials that came into contact with the tritiated methane were used in XENON1T.

[4] Activation underground is negligible due to the low muon flux.

This is several orders of magnitude above the 'measured' value of $\sim 10^{-25}$, but several removal processes need to be accounted for. Given the measured $\sim$ ppmv water concentration in the xenon gas inventory [110] and equilibrium conditions [114], this cosmogenically-activated tritium will predominantly take the form of HTO and react chemically in a similar way to water vapor.[5] Once the xenon inventory was brought underground, HTO from cosmogenic activation would be removed by (1) radioactive decay (2) adsorption onto the walls of the large storage vessel (known as ReStoX [30]), and (3) purification via a high-efficiency getter, where a hydrogen removal unit (HRU) is employed specifically to remove hydrogen and its isotopes [30, 88].

The xenon gas was underground for ~2 years before the operation of XENON1T, meaning only a small amount of tritium would have decayed before data-taking. However, we expect tritiated water to be removed by many orders of magnitude from steps (2) and (3) above. For (2), the transfer of the xenon from room temperature, high-pressure bottles to the ReStoX cryogenic storage system would have resulted in a large amount of HTO vapor condensing on the walls of the cold storage vessel. We estimate that this reduction factor is at least ~4000 (G. Plante, Personal communication). When filling the XENON1T TPC, the xenon is passed through a hot zirconium getter, which removes water from the xenon gas with an efficiency >99.99%, a further reduction of 4 orders of magnitude. Combining steps (2) and (3), we thus conclude that a tritium contribution from cosmogenic activation would have been reduced to a maximum level of $\sim 10^{-27}$, far too small to account for the excess.

Tritium may also be introduced as HTO and HT via their respective atmospheric abundances. Water and hydrogen, and therefore tritium, may be stored inside materials, such as the TPC reflectors and the stainless steel of the cryostat. These sources of tritium are expected to emanate from detector and subsystem materials, along with its non-tritiated counterparts, at a rate in equilibrium with their removal via getter purification. Tritium can be found in water at a concentration of $(5-10) \times 10^{-18}$ atoms of ^{3}H for each atom of hydrogen in H_2O [113, 135, 136]. Here we assume the same abundance of ^{3}H in atmospheric H_2 as for water.[6] Using the best-fit rate of tritium and the HTO atmospheric abundance, we require a combined $(H_2O + H_2)$ impurity concentration C_H of

$$\begin{aligned} C_H \simeq & \left(\frac{6.2 \times 10^{-25}\ \text{T atoms}}{\text{Xe atom}}\right)\left(\frac{\text{H atom}}{(5{-}10) \times 10^{-18}\ \text{T atoms}}\right) \\ & \times \left(\frac{(\text{H}_2\text{or}\text{H}_2\text{O})\ \text{molecule}}{2\ \text{H atoms}}\right) \\ & \simeq (30{-}60)\ \text{ppb (mol)} \end{aligned}$$

[5] With the caveat that there is some different radiochemistry expected to an extent.

[6] Although geographical and temporal HT abundances in the atmosphere vary due to anthropogenic activities, HT that reaches the Earth's surface undergoes exchange to HTO within 5 h [106, 127].

in the LXe target to make up the excess. This concentration could be composed of a combination of H_2O and H_2.

Since water impurities affect optical transparency, the high light yield in SR1 indicates an $O(1)$-ppb H_2O concentration [23, 25]. Therefore, the light yield measured in XENON1T excludes HTO as the sole contribution to the required C_H above, though it could be contributing in a sub-dominant way.

Unlike the HTO case, we cannot constrain the HT concentration. To calculate the amount of HT in the detector, we would need a measurement on the H_2 concentration in the liquid xenon that we do not have. Nonetheless, we can qualitatively compare to concentration of other impurities in the detector. Given an electron lifetime of ~650 μs (at 81 V/cm) [31] we can conclude that the O_2-equivalent, electronegative impurities must have reached sub-ppb levels in SR1 [24]. Therefore a factor ~100 higher H_2 concentration than that of electronegative impurities is required in order to have a tritium concentration consistent with the observed excess. We find this unlikely, but we cannot rule it out at this time.

In conclusion, a tritium contamination from either cosmogneic activation or HTO are both ruled out as explanations for the excess, but we cannot exclude a possible HT contamination of the level required. Additionally, we note that various factors contribute further to the uncertainty in estimating a tritium concentration within a LXe environment, such as its unknown solubility and diffusion properties, as well as the possibility that it may form molecules other than HT and HTO. Since the information and measurements necessary to quantify the tritium concentration are not available, we can neither confirm nor exclude it as a background component. Therefore, we treat trace amounts of tritium as an additional hypothesis. It is thus useful to compare the $B_0 +^3$ H to the B_0-only model, as already shown in Fig. 4.10. Comparing the likelihood ratio of the two models, the B_0-only hypothesis is rejected at 3.2σ.

In the next section, we consider the three new-physics channels discussed in Chap. 1. Since the tritium hypothesis can neither be confirmed nor excluded, we will report the main results using the B_0-only background model, and describe how things change when including tritium as a background component.

4.4 Searches for New Physics in XENON1T ER Data

Now we turn to the main scientific goal of this analysis, the search for new physics in the form of solar axions, neutrino magnetic moment, and bosonic dark matter. The excess is relevant for all three of these signal models, along with the tritium hypothesis. Since we cannot confirm tritium to be present at the level required to explain the excess, our main results use the main background model B_0. For the axion and neutrino magnetic moment analyses, we also report how the results change if tritium were included as an unconstrained background component.

The outline of the next several sections is as follows. First, we consider the solar axion signal model. As a low-energy signal, it can be used to explain the excess rather well. Using the three dimensional statistics framework discussed in Sect. 3.3, we report a three-dimensional confidence volume for the axion couplings to matter. We then discuss these constraints in the context of other experiments and astrophysical analyses. While the solar axion results are by design model-independent, we consider the results in the context of benchmark QCD axion models DFSZ and KSVZ. Finally, based on the tritium discussion above, we report how the results change if tritium were added to the background model.

Section 4.6 reports on the search for an enhanced neutrino magnetic moment. Also a low-energy signal, this hypothesis also describes the observed excess well. We report a confidence interval on the value of μ_ν, and discuss implications relating to other constraints, including a cross-check using another analysis with XENON1T in which only ionization signals are considered. We also compare this signal model to the tritium hypothesis.

The bosonic dark matter results are reported in Sect. 4.7, where world-leading constraints are made on the couplings for these particles. Low-mass bosonic dark matter also describes the low-energy excess well, and what mass gives the best fit.

4.5 Search for Solar Axions

4.5.1 Main Results

We search for ABC, ^{57}Fe, and Primakoff axions simultaneously by including a 3-component axion signal component when fitting the data. The best-fit of the B_0 + axion model compared to the B_0-only model is shown in Fig. 4.11, where better agreement is observed for the axion model. The likelihood ratio between the alternate (B_0 + axion) and the null (B_0) hypothesis is $q_0 = 14.5$. Comparing the observed q_0 to the null distribution constructed in Sect. 3.3, a result as extreme or more so than the SR1 value is expected to happen by chance in only 0.03% of

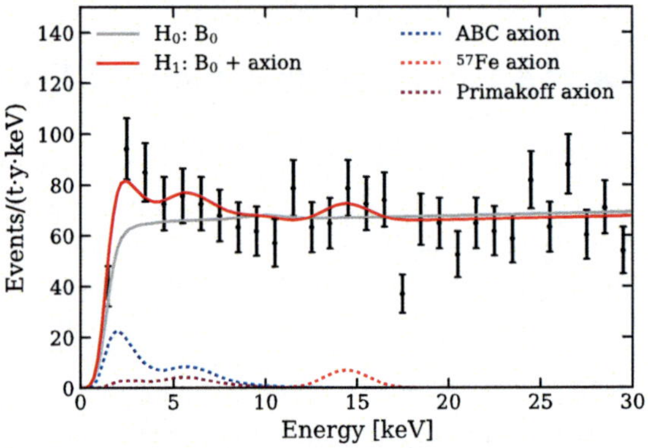

Fig. 4.11 Fit results for the axion hypothesis (red) compared to the B_0-only hypothesis (gray)

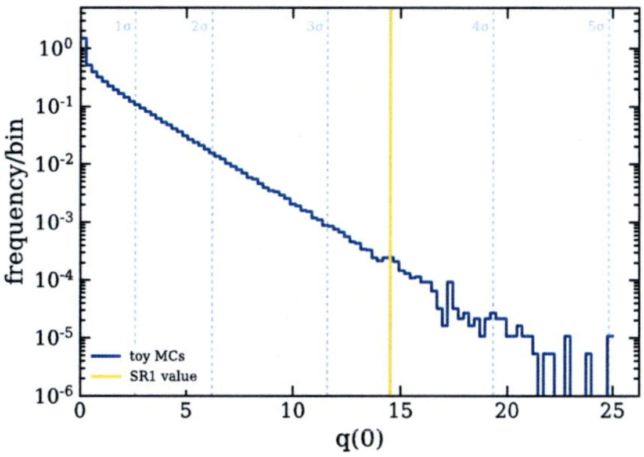

Fig. 4.12 Comparison of the observed SR1 likelihood ratio q_0 and the null distribution calculated in Sect. 3.3 with toyMC methods. This implies a rejection of the null hypothesis at 3.4σ

experiments. That is, the B_0-only hypothesis rejected at 3.4σ. The comparison of the SR1 value to the null distribution is shown in Fig. 4.12 for clarity.

A three-dimensional confidence volume (90% C.L.) was calculated in the space of g_{ae} vs. $g_{\mathrm{ae}}g_{\mathrm{a}\gamma}$ vs. $g_{\mathrm{ae}}g_{\mathrm{an}}^{\mathrm{eff}}$ using the method described in Sect. 3.3. In reality, the confidence volume was first calculated in the space of the axion component rates; i.e., R_{ABC} vs. $R_{\mathrm{Primakoff}}$ vs. R_{Fe57}. Recall from Sect. 3.3 that the edges of the *rate* confidence volume are defined by the intersection of our three-dimensional test statistic $q(\mathrm{ABC}, \mathrm{Primakoff}, {}^{57}\mathrm{Fe})$ with the critical volume calculated shown in Fig. 3.15. The result of this procedure is shown in Fig. 4.13. This volume is inscribed in the cuboid given by

$$g_{\mathrm{ae}} < 3.7 \times 10^{-12}$$

$$g_{\mathrm{ae}}g_{\mathrm{an}}^{\mathrm{eff}} < 4.6 \times 10^{-18}$$

$$g_{\mathrm{ae}}g_{\mathrm{a}\gamma} < 7.6 \times 10^{-22}\ \mathrm{GeV}^{-1}.$$

The cuboid is perhaps easier to visualize and work with than the confidence volume it encloses, but it is also more conservative (it displays over-coverage) and does not describe the correlations between the parameters. Further visualization of the confidence volume can be found in Fig. 4.14, which shows its two-dimensional projections. For the ABC–Primakoff and ABC–^{57}Fe projections (Fig. 4.14 top and middle, respectively), g_{ae} can be easily factored out of the y-axis to plot $g_{\mathrm{a}\gamma}$ vs g_{ae} (top) and $g_{\mathrm{an}}^{\mathrm{eff}}$ vs g_{ae} (middle). This is not as straightforward for the ^{57}Fe-Primakoff projection (Fig. 4.14 bottom); hence we leave this plot in the space of $g_{a\gamma}g_{ae}$ vs g_{an}^{eff}. Also shown in Fig. 4.14 are constraints from other axion searches [11, 48, 53, 59, 95, 102, 155] as well as predicted values from the benchmark QCD models DFSZ and KSVZ.

Figure 4.14 (top) is extracted from the projection onto the ABC–Primakoff plane. Since the ABC and Primakoff components are both low-energy signals, the 90% confidence region is anti-correlated in this space and—due to the presence of

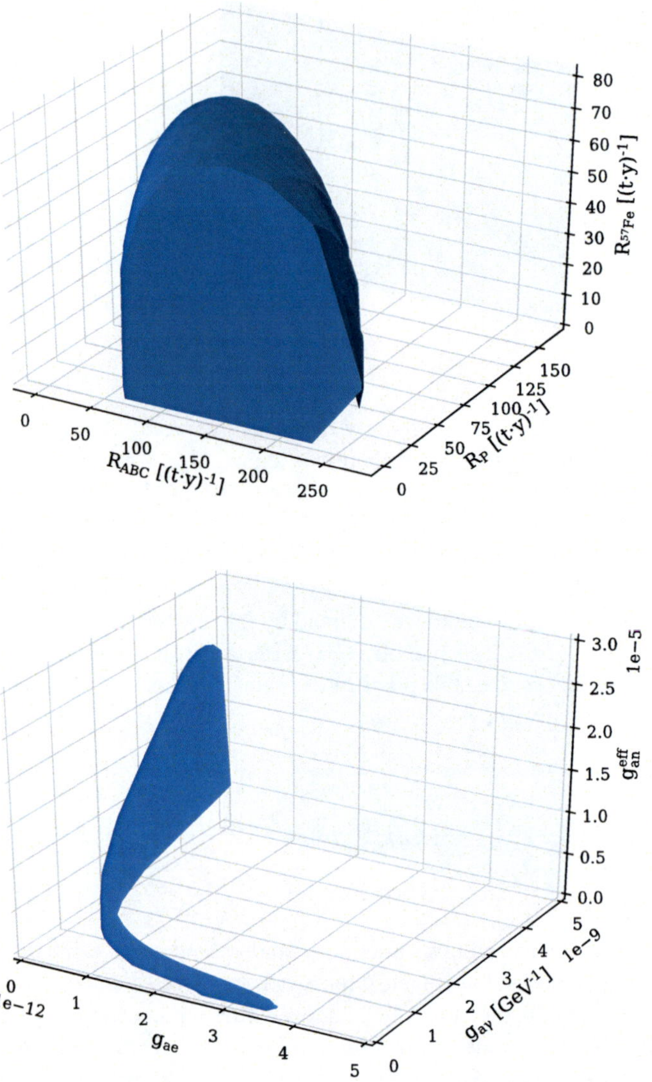

Fig. 4.13 Three-dimensional confidence volume for the solar axion analysis. Top: volume defined in the space of the three axion component rates. Bottom: after converting to the space of the axion couplings. Though not obvious from the plot due to the linear scale, as $g_{ae} \rightarrow 0$, both g_{an}^{eff} and $g_{a\gamma}$ $\rightarrow \infty$

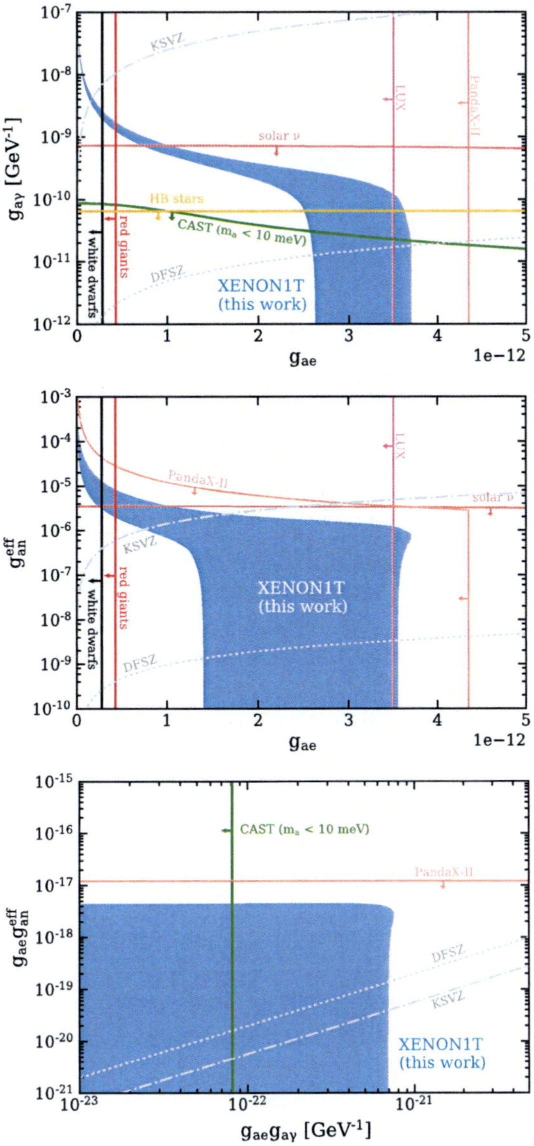

Fig. 4.14 Constraints on the axion-electron g_{ae}, axion-photon $g_{\mathrm{a}\gamma}$, and effective axion-nucleon $g_{\mathrm{an}}^{\mathrm{eff}}$ couplings from a search for solar axions. The shaded blue regions show the two-dimensional projections of the three-dimensional confidence surface (90% C.L.) of this work, and hold for $m_{\mathrm{a}} < 100\,\mathrm{eV/c^2}$

the low-energy excess—suggests either a non-zero ABC component or non-zero Primakoff component. Since our result gives no absolute lower bound on g_{ae}, the limit on the product $g_{\mathrm{ae}}g_{\mathrm{a}\gamma}$ cannot be converted into a limit on $g_{\mathrm{a}\gamma}$ on its own; i.e., with $g_{\mathrm{ae}}g_{\mathrm{a}\gamma}$=$7.6 \times 10^{-22}\,\mathrm{GeV}^{-1}$, $g_{\mathrm{a}\gamma} \rightarrow \infty$ as $g_{\mathrm{ae}} \rightarrow 0$, as shown in Fig. 4.14 (top).

Figure 4.14 (middle) is taken from the projection onto the ABC–^{57}Fe plane. Unlike the ABC-Primakoff case, these two signals are not degenerate; however,

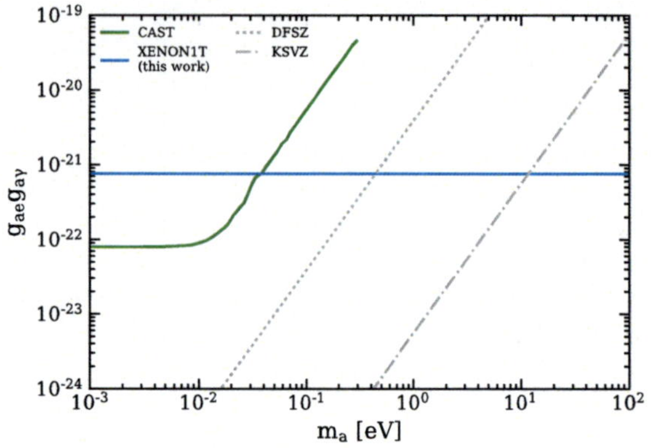

Fig. 4.15 Limit on the product $g_{ae}g_{a\gamma}$ as a function of axion mass m_a from this work, along with that from CAST [53]. Since the CAST detection mechanism depends on m_a, they lose sensitivity to high-mass axions

they still display anti-correlated behavior. The reason for this is that the test statistic q is relatively large with small g_{ae}, meaning small changes in the ^{57}Fe rate about the best-fit make q cross the 90% threshold value and thus be excluded by our 90% confidence surface. There is no statistical significance ($<1\,\sigma$) for the presence of a 14.4 keV peak from ^{57}Fe axions.

Lastly, Fig. 4.14 (bottom) shows the projection onto the Primakoff-^{57}Fe plane, where no correlation is observed. The Primakoff and ^{57}Fe components are both allowed to be absent as long as there is a non-zero ABC component. This means that, of the three axion signals considered, the ABC component is the most consistent with the observed excess.

To summarize, interpreting the XENON1T data under a solar axion signal model, the resulting 90% confidence volume suggest either a non-zero ABC component or a non-zero Primakoff component. However, the coupling values needed to explain this excess are in strong tension with stellar cooling constraints [48, 59, 81, 101, 155]. Part of the volume is also in tension with constraints from CAST [53]; note, however, that the CAST constraints as shown are valid only for axion masses below 10 meV/c^2, while those from XENON1T and similar experiments hold for all axion masses up to ~100 eV/c^2. The CAST experiment loses sensitivity to high-mass axions since they search for X-rays produced from the inverse Primakoff effect in the presence of an external magnetic field. To explicitly show this dependence on the mass, we also plot in Fig. 4.15 our solar axion limit in the space $g_{ae}g_{a\gamma}$ vs m_a, the natural space for the CAST result. Since the XENON1T detection mechanism has no dependence on m_a (after assuming $m_a < 100$ eV), our limit is a flat line in this space. The CAST limit, on the other hand, does depend on mass. The upper limit from XENON1T is more stringent than that from CAST for masses above ~40 meV.

4.5.2 Interpretation Under Specific QCD Models

We consider the model-independent confidence volume the main result for the solar axion analysis. Nonetheless, it is interesting to interpret our result assuming the QCD axion models described in Sect. 1.2.1, the DFSZ (I and II) and hadronic/KSVZ models. Recall that the DFSZ models have well-defined electromagnetic anomaly E ($E = 8$ for DFSZ-I, $E = 2$ for DFSZ-II) but an additional free parameter $\beta_{\rm DFSZ}$, while hadronic models have a range of possible E values but no analogous β parameter. For each model, our goal is to find what values of m_a, if any, are consistent with the XENON1T confidence volume, along with the respective $\beta_{\rm DFSZ}$ or E values.

DFSZ We first consider the DFSZ class of models, elegantly referred to as DFSZ-I and DFSZ-II. These models are essentially identical apart from having different values of the electromagnetic anomaly E, and so we describe the approach for DFSZ-I only and quote the results for both.

Recall from Sect. 1.2.1 that, for a DFSZ model,

$$g_{\rm ae} = \frac{m_{\rm e}}{3 f_{\rm a}} \cos^2 \beta_{\rm DFSZ}, \tag{4.1}$$

$$g_{\rm a\gamma} = \frac{\alpha}{2\pi f_{\rm a}} \left(\frac{E}{N} - \frac{2}{3} \frac{4+z}{1+z} \right), \tag{4.2}$$

$$g_{an}^{\rm eff} = -1.19 g_{an}^0 + g_{an}^3, \tag{4.3}$$

where g_{an}^0 and g_{an}^3, which are just combinations of g_{an} and g_{ap}, are related to $\beta_{\rm DFSZ}$. To be explicit,

$$\begin{aligned} g_{an}^{\rm eff} &= \frac{-1.19}{2}(g_{ap} + g_{an}) + \frac{1}{2}(g_{ap} - g_{an}) \\ &= -\frac{1}{2}(0.19 g_{ap} + 2.19 g_{an}) \\ &= -\frac{m_p}{2 f_a} \Big[0.19(-0.47 + 0.88 C_u - 0.39 C_d) \\ &\quad + 2.19(-0.02 + 0.88 C_d - 0.39 C_u) \Big] \\ &= \frac{m_p}{2 f_a} \left(0.089 + \frac{0.69}{3} \sin^2 \beta_{\rm DFSZ} - \frac{1.85}{3} \cos^2 \beta_{\rm DFSZ} \right), \end{aligned} \tag{4.4}$$

where we have taken $m_p \approx m_n$ and used the DFSZ values for C_u and C_d. We use these relations and our model independent confidence volume to make statements on m_a, $g_{a\gamma}$, g_{ae}, g_{ap}, and g_{an} within the context of a DFSZ model.

In Fig. 4.14 (top) we show the projection of the confidence volume in $g_{a\gamma}$-vs-g_{ae} space. This plot can be converted to the space of g_{ae}-vs-m_a space by recognizing that, for the DFSZ-I model ($E = 8$), there is a one-to-one correspondence between $g_{a\gamma}$ and m_a via Eq. (4.2). After this transformation, Eq. (4.1) can then be used to find what values of m_a and β_{DFSZ} are consistent with the (transformed) projection.

The procedure above gives a range of m_a and β_{DFSZ} consistent with the g_{ae}-vs-$g_{a\gamma}$ projection, but we can go one step further by considering our constraint on g_{an}^{eff}. For this we can use Fig. 4.14 (middle), where we plot g_{an}^{eff} vs g_{ae}. Since these two variables are (non-trivially) related via β_{DFSZ}, this plot can be numerically converted to the space of β_{DFSZ} vs g_{ae}. But these two variables themselves are in turn related to m_a via Eq. (4.1), allowing us to transform to g_{ae}-vs.-m_a for this projection as well. Note that there is no need to repeat a similar procedure for Fig. 4.14 (bottom) due to the fact that we have used the constraints on all three couplings already.

This procedure uses the model-specific relationships amongst $g_{a\gamma}$, g_{ae}, and g_{an}^{eff} to give two consistent regions in the space of g_{ae} vs m_a. The intersection of these two regions gives the final result: a model-dependent "consistence region" that overlaps with the XENON1T confidence volume, which is shown in Fig. 4.16 (top). We refer to this as a consistence region, rather than a *confidence* region, due to the fact that our statistical framework was developed to report results in a model-independent fashion; hence, the confidence volume is not guaranteed to have the proper coverage after including model-dependent assumptions *ex post facto*. That is, these results do not have well-defined coverage in the statistical sense, but are instead just the models/values that intersect our confidence volume.

The consistence regions for the DFSZ models give a mass range of $m_a \sim 0.1$–4.1 eV for the DFSZ-I model and $m_a \sim 0.1$–2.3 eV for DFSZ-II. This mass range can then be used to plot the consistence region in the space of $g_{a\gamma}$-vs-m_a, again due to the one-to-one correspondence in Eq. (4.2). Further, using this mass range, the g_{ae}-vs.-m_a result, and Eq. (4.4), we can plot the consistence region in the space of g_{an}-vs-g_{ap}. These regions are all summarized in Fig. 4.16 and compared to other constraints.

Figure 4.16 contains a lot of information that we summarize here. The top figure (g_{ae} vs m_a) shows that the XENON1T result could be consistent with a $m_a \sim O(0.1 - 1)$ eV DFSZ-type axion with $g_{ae} \sim 2 - 3 \times 10^{-12}$. This is not ruled out by any direct measurements; however, as also emphasized in the main result, such a large axion-electron coupling is in strong tension with (indirect) astrophysical analyses of red giants and white dwarfs.

The middle figure ($g_{a\gamma}$ vs m_a) gives the range of $g_{a\gamma}$ values expected for DFSZ-type axions consistent with the XENON1T result. In this way we can see if an $O(0.1 - 1)$ eV DFSZ axion is ruled out due to its photon coupling. Similarly to the top figure, a DFSZ interpretation of the XENON1T result is not in tension with any direct experiment; even CAST, which loses starts to lose detection sensitivity around $O(10)$ meV, does not rule out any of the consistence region. The strongest indirect constraint in this region comes from a study of Horizontal Branch (HB) stars [48]. This precludes part of the consistence region, but not all of it; for $m_a <$ 480 (270) meV, the XENON1T DFSZ region is not excluded by any $g_{a\gamma}$ constraint.

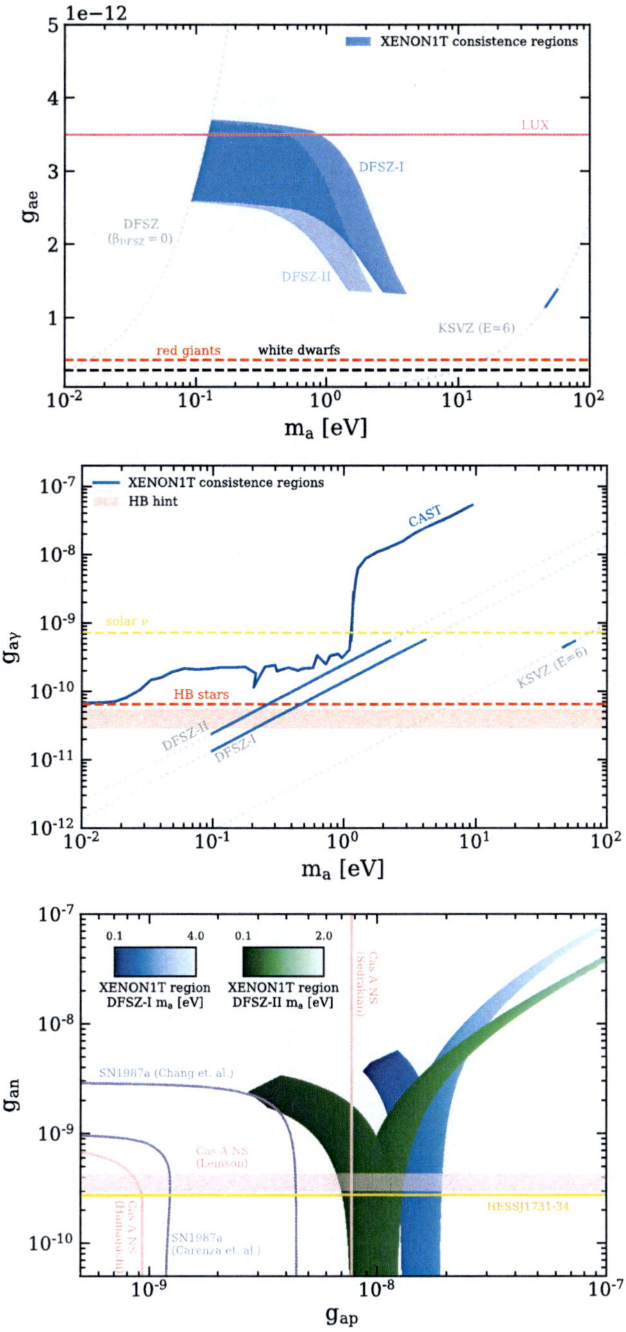

Fig. 4.16 Interpreting the XENON1T excess as a solar axion signal under DFSZ-I, DFSZ-II, and KSVZ models. Consistent regions are shown in light blue and/or green. For KSVZ models, the only consistent value of the electromagnetic anomaly is $E = 6$, which falls well outside the axis limits of the bottom figure

Interestingly, the same HB analysis reported excess cooling, which, if attributed to axions, would suggest a $g_{a\gamma} = 0.45^{+0.12}_{-0.16} \times 10^{-10}$ GeV^{-1}, consistent with the XENON1T DFSZ interpetation.

Figure 4.16 (bottom) shows the consistence region in the space of g_{an} vs g_{ap}, with he mass information included on the color scale. If from a DFSZ axion, the XENON1T result would suggest a $g_{ap} \sim 2 \times 10^{-9}$–$10^{-8}$ and $g_{an} \lesssim 10^{-8}$, with precise values depending on the specific DFSZ model, m_a, and β_{DFSZ}. There is no reported direct constraint in this parameter space, so we compare to indirect results from SN1987a and neutron stars (NS). Over the years many groups have reported constraints the axion couplings with SN1987a, sometimes with differences of up to an order of magnitude in the final results. We include two such results here [66, 68], both of which suggest small tension with the XENON1T result. Uncertainties on supernova collapse are also a concern, so these constraints are not considered sharp bounds [94, 152].

The neutron star constraints are based on rates of cooling, both in the Cassiopeia A (Cas A) supernova remnant and the HESS J1732 NS. Similarly to the SN1987a case, there have been many studies of Cas A in the context of axions, each with slightly different results [104, 123, 148, 152]. Reference [123] suggests a possible axion hint for $g_{an} \sim 4 \times 10^{-10}$, and [148] gives an upper limit on $g_{ap} \lesssim 7.7 \times 10^{-9}$, both of which would be consistent with a DFSZ interpration of our result. The most stringent limit using Cas A comes from [104], whose result suggest moderate tension with the XENON1T result. However, the conclusion from [104] is subject to a large uncertainty on the stellar envelope parameter that would change the results for DFSZ axions substantially;[7] in fact, if the stellar envelope is maximally carbon-rich, the authors of [104] suggest that the Cas A rate of cooling could actually be consistent with a DFSZ axion of $f_a \sim 10^{-8}$ GeV ($m_a \sim 60$ meV). Lastly, the HESS J1732 NS result [61], based on a different neutron star, gives a strict upper limit of $g_{an} \lesssim 3 \times 10^{-10}$, which would be consistent with the XENON1T result for $g_{ap} \sim 10^{-8}$. To summarize, some astrophysical analyses constraining the axion-nucleon couplings are consistent with the XENON1T result, while others—even though they are often based on the same astrophysical objects—suggest small to moderate tension.

KSVZ For hadronic/KSVZ models, recall that for a given value of E there is a one-to-one correspondence between m_a and all three axion couplings considered here; there is no parameter analogous to β_{DFSZ}. This simplifies the procedure. We again use the top and middle projection of Fig. 4.14, but in this case we can directly convert both of these projections into mass ranges for a given value of E. If these two mass ranges overlap at all, we conclude that E and the resulting overlapping mass range is consistent with our confidence volume. We then loop over all values of E between 0 and 170, and see how many consistent models there are. Only one

[7] They claim the limit for KSVZ axions is robust against the uncertainty due to the small g_{an}, while this is not the case for DFSZ axions.

value of E is consistent with the XENON1T result: $E = 6$. The corresponding consistent mass range for this model is $m_a \sim 46 - 56$ eV and is shown in Fig. 4.16.

We also include the KSVZ interpretation In Fig. 4.16 (top and middle). As discussed, stellar cooling constraints on white dwarfs and red giants are in strong tension with the XENON1T result. Under a KSVZ interpretation, the consistent range of $g_{a\gamma}$ is also in strong tension with the constraint from HB stars. The KSVZ result, omitted from the bottom as it falls far outside the axis window, is in very strong tension with constraints from both SN1987a and neutron stars.

To conclude our discussion on QCD interpretations, we find that a DFSZ axion with mass between 100 meV and 4.1 eV would be consistent with the XENON1T result. Such an axion is not ruled out by any direct measurements, though is in tension with g_{ae} limits on white dwarfs and red giants, and in loose tension with SN1987a and NS constraints on g_{ap} and g_{an}, respectively. It is also worth noting that a DFSZ axion of $O(100)$ meV is consistent with some hints of possible axion signals in HB stars as well as neutron stars. The KSVZ model does not agree very well with the XENON1T result, but there is a thin sliver of overlap for $E = 6$ and $m_a \sim 46 - 56$ eV. Such an axion is in strong tension with nearly all astrophysical constraints discussed.

4.5.3 Including Tritium as a Background

As described in Sect. 4.3.3, we cannot exclude tritium as an explanation for this excess. Thus, we report on an additional statistical test, where an unconstrained tritium component was added to the background model B_0 and profiled over alongside the other nuisance parameters. In this case, the null hypothesis is the background model plus tritium ($B_0 + {}^3$H) and the alternative includes the three axion signal components ($B_0 + {}^3$H + axion), where tritium is unconstrained in both cases.

Adding an unconstrained tritium component complicates the statistical inference slightly. Since we have no prediction/constraint for the concentration, the distribution of the test statistic under the null hypothesis is not well defined a priori; i.e., only after looking at the data do we have an estimate of a possible tritium background. We thus treat the inference with a toy-MC approach, similarly to the standard axion inference, whereas this time we simulate many $B_0 +^3$ H datasets— where the mean injected tritium signal is given by the SR1 best-fit value— and for each one compare the likelihood ratio of $B_0 +^3$ H and $B_0 +^3$ H + axion, with tritium unconstrained in both cases. This is a subtle but important point: for each simulation the null hypothesis is given by the unconstrained tritium fit, not the true injected tritium signal. We do it this way to account for the fact that we use the data to decide what the best tritium concentration is, not outside knowledge as is the case for other background components.

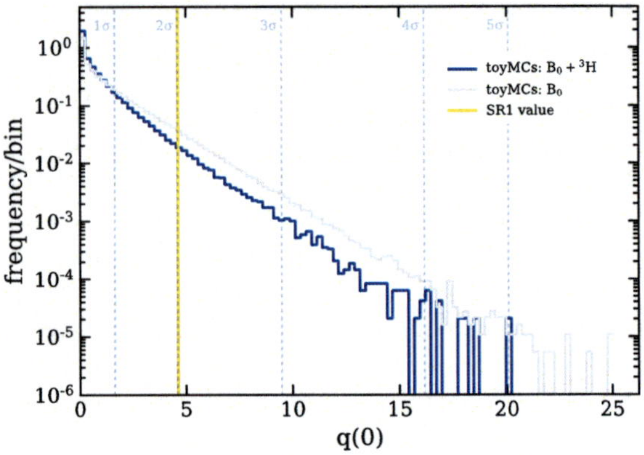

Fig. 4.17 Distribution of the test statistic under null hypothesis $B_0 +^3$ H (blue), where tritium is unconstrained. Also shown for comparison is the null distribution of the B_0 hypothesis (gray) and the value of q_0 for SR1, giving a 2.0σ significance of rejecting the $B_0 +^3$ H hypothesis when compared to the $B_0 +^3$ H + axion one

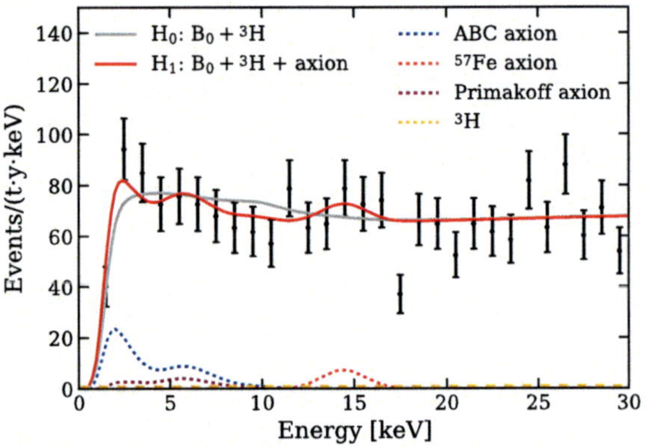

Fig. 4.18 Fit results for the $B_0 +^3$ H + axion hypothesis (red) compared to the $B_0 +^3$ H hypothesis (gray)

By leaving tritium unconstrained the null hypothesis gains additional freedom, which squeezes our null test statistic distribution to lower values. That is, by letting the ^{3}H component float free in the fit for the null hypothesis, it becomes more unlikely that the alternate ^{3}H + axion hypothesis would give a much better fit to the data. This is shown in Fig. 4.17, where one can see the $B_0 +^3$ H distribution is squeezed relative to the to B_0-only one.

When we compare these two hypotheses, the null $B_0 +^3$ H vs. the alternate $B_0 +^3$ H + axion, for the SR1 data, we find that the null hypothesis is disfavored at 2.0σ. The SR1 test statistic is included in Fig. 4.17 for comparison with the null distribution, and the fits for this analysis are shown in Fig. 4.18. The tritium component is negligible in the alternate best-fit, but its presence allows for a better fit under—and thus a reduced significance of rejecting—the null hypothesis.

4.6 Search for an Enhanced Neutrino Magnetic Moment

4.6.1 Main Result

We now turn to the search for an enhanced neutrino magnetic moment, which similarly to the axion signal would express itself as an excess at low energies. The best-fit under the $(B_0 + \mu_\nu)$ model is shown in Fig. 4.19, along with the B_0-only fit as before. In this case, the likelihood ratio test statistic q_0 is 10.0. Assuming the test statistic is asymptotically distributed, this implies that the B_0-only hypothesis is rejected at 3.2σ.

Since the discovery significance exceeds 3σ, we report a two-sided 90% confidence interval, given by

$$\mu_\nu \in [1.4,\ 2.9] \times 10^{-11}\, \mu_B,$$

and shown in Fig. 4.20 along with the constraints from other searches. The upper boundary of this interval is very close to the limit reported by Borexino [7], which is currently the most stringent direct detection constraint on the neutrino magnetic moment. Similar to the solar axion analysis, if we infer the excess as a neutrino magnetic moment signal, our result is in strong tension with indirect constraints from analyses of white dwarfs [71] and globular clusters [90]. It is important to note that the neutrino flavor does impact the interaction involving the magnetic moment, which in reality is a 3×3 matrix due to neutrino mixing. Our result, based on a flavor-insensitive detection of solar neutrinos, is thus directly comparable to Borexino's, but not necessarily to Gemma's (reactor electron anti-neutrinos) or the astrophysical limits (electron neutrinos).

As a cross-check, we also searched for an enhanced neutrino magnetic moment using an S2-only analysis, where the requirement of an S1 signal is dropped entirely. This type of analysis allows for a lower energy threshold of ~200 eV; however, lacking a complete background model, such an analysis has no discovery potential

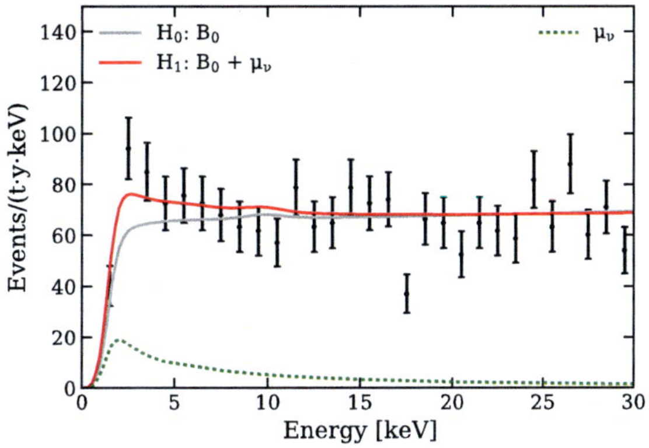

Fig. 4.19 Fit results for the magnetic moment hypothesis (red) compared to the B_0-only hypothesis (gray)

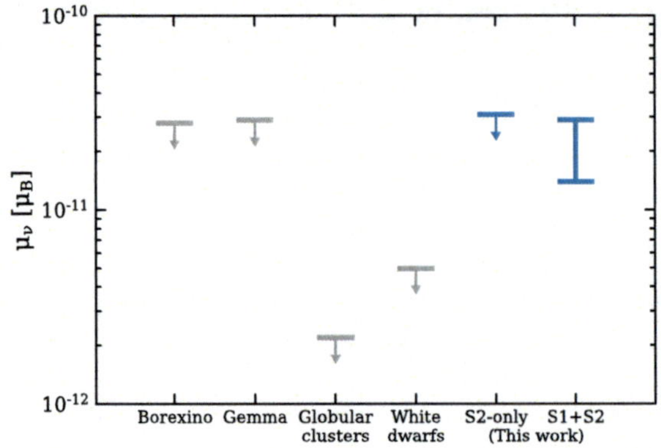

Fig. 4.20 Constraints (90% C.L.) on the neutrino magnetic moment from this work compared to experiments Borexino [7] and Gemma [56], along with astrophysical limits from the cooling of globular clusters [90] and white dwarfs [71]. The constraint from XENON1T using ionization signal only (S2-only) is also shown. Arrows denote allowed regions. The upper boundary of the interval from this work is about the same as that from Borexino and Gemma. If we interpret the low-energy excess as a neutrino magnetic moment signal, its 90% confidence interval is in strong tension with the astrophysical constraints

and can only be used to place upper limits. The XENON1T collaboration recently published the results of its main S2-only analysis [33], which included world-leading constraints on light dark matter, among others. Since the expected neutrino magnetic moment spectrum grows exponentially with decreasing energy, an S2-only analysis is a useful cross-check of the magnetic moment hypothesis.[8] We employed the XENON1T S2-only analysis to place an upper limit on the neutrino magnetic moment of $\mu_\nu < 3.1 \times 10^{-11}\ \mu_B$ (90% C.L.), which is also included in Fig. 4.20. Since this limit is larger than the upper limit of our 2-sided interval above, the S2-only analysis is consistent with a neutrino magnetic interpretation of the XENON1T excess.

This is the most sensitive search to date for an enhanced neutrino magnetic moment with a dark matter detector, and suggests that this beyond-the-SM signal be included in the physics reach of other dark matter experiments.

4.6.2 Including Tritium as a Background

As in Sect. 4.5, we report on an additional statistical test where an unconstrained tritium component was included in both null and alternative hypotheses. In this

[8] We did perform the same analysis for the axion and tritium hypotheses, but the spectra of these two are almost entirely above 1 keV, which makes the S2-only analysis much less sensitive and thus less relevant. The upper limits of the S2-only analysis for these two are more than an order of magnitude above the results presented in Sects. 4.3.3 and 4.5.

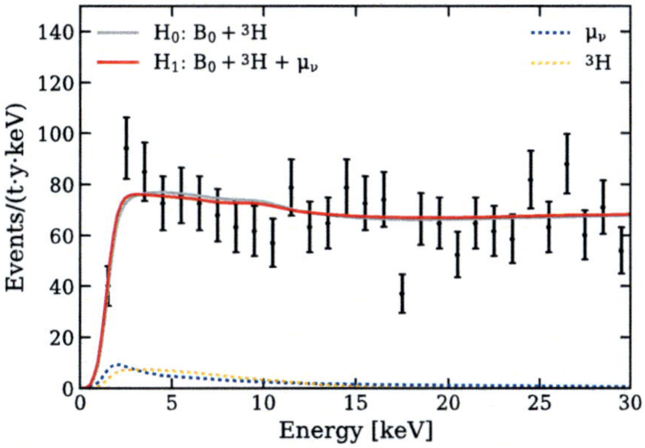

Fig. 4.21 Fit results for the B_0+^{3}H+μ_ν hypothesis (red) compared to the B_0+^{3}H hypothesis (gray)

test the best fit includes a mix of magnetic moment and tritium spectra, as seen in Fig. 4.21. Due to the similar spectra of the magnetic moment and tritium models, the significance of the neutrino magnetic moment signal is reduced to 0.9σ. Thus, contrary to the solar axion case, the magnetic moment hypothesis is not favored over the tritium hypothesis based on the observed energy spectrum.

4.7 Bosonic Dark Matter Results

For the bosonic dark matter search, we iterate over masses in 1-keV steps between 1 and 210 keV/c^2 to search for peak-like excesses. This iteration procedure introduces a trial factor for this analysis, which needs to be accounted for when reporting the global significance [103]. We calculated the trial factor using toy Monte Carlo methods, in which we simulated many background-only datasets, searched over our pre-selected masses, and saved the largest local significance, the distribution of which is used for reporting the global significance.

When looping over the pre-defined masses in 1 keV steps, we find no global significance over 3σ for this search under the background model B_0. We thus set an upper limit on the couplings g_{ae} and κ as a function of particle mass. These upper limits (90% C.L.) are shown in Fig. 4.22, along with the sensitivity band in green (1σ) and yellow (2σ). The losses of sensitivity at 41.5 and 164 keV are due to the ^{83m}Kr and ^{131m}Xe backgrounds, respectively, and the gains in sensitivity at around 5 and 35 keV are due to increases in the photoelectric cross-section in xenon. The lack of smoothness is due to a combination of the photoelectric cross-section, the logarithmic scaling, and the fact that the energy spectra differ significantly across the range of masses. For most masses considered, XENON1T sets the most stringent direct-detection limits to date on pseudoscalar and vector bosonic dark matter couplings.

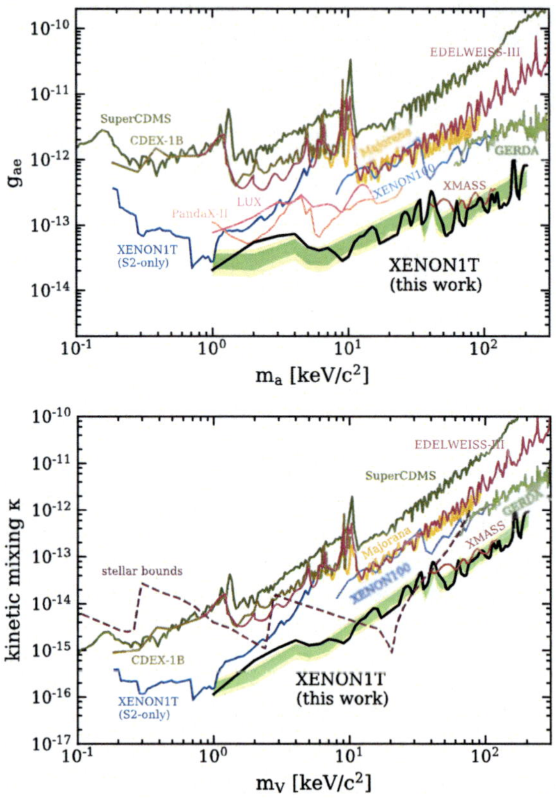

Fig. 4.22 Constraints on couplings for bosonic pseudoscalar ALP (top) and vector (bottom) dark matter, as a function of particle mass. The XENON1T limits (90% C.L.) are shown in black with the expected 1 (2) σ sensitivities in green (yellow). Limits from other detectors or astrophysical constraints are also shown for both the pseudoscalar and vector cases [4, 5, 8, 11, 19, 29, 29, 33, 43, 47, 95, 157]

Due to the presence of the excess, we performed an additional fit using the bosonic dark matter signal model, with the particle mass allowed to vary freely between 1.7–3.3 keV. The result gives a favored mass value of $(2.3 \pm 0.2)\,\text{keV/c}^2$ (68% C.L.) with a 3.0σ global (4.0σ local) significance over background. Since this mass point was not included in the pre-defined masses to loop over, an upper limit was not extracted for this datapoint.[9] A log-likelihood ratio curve as a function of mass is shown in Fig. 4.23 (left), along with the asymptotic 1-σ uncertainty. The spectral fit of the 2.3 keV peak is illustrated in Fig. 4.23 (right).

[9] The global significance study also did not consider this mass point, so the global significance is likely in reality slightly less than 3σ.

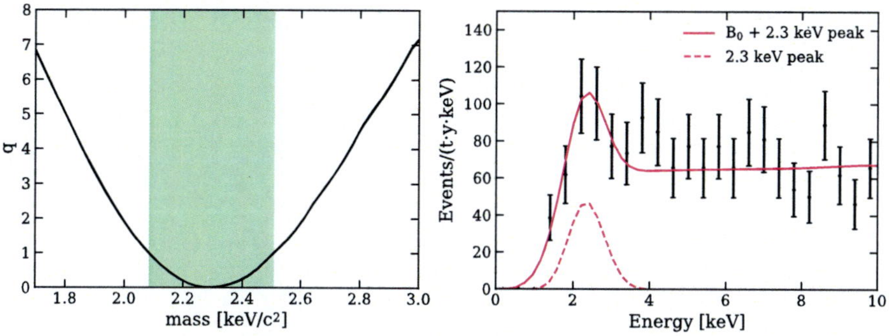

Fig. 4.23 Left: the log-likelihood ratio for different bosonic dark matter masses with respect to the best-fit mass at 2.3 keV/c^2. At each mass, we show the result for the best-fit coupling at that mass. The green band shows an asymptotic 68% C.L. confidence interval on the bosonic dark matter mass. As noted in the text, the global significance for this model is 3.0 σ. Right: best-fit of a 2.3 keV peak and B_0 to the data. A 0.4 keV binning is used for better visualization

The best-fit value of a 2.3 keV monoenergetic peak is quite close to that expected from a potential ^{37}Ar background (indeed this motivated the argon hypothesis). While the ^{37}Ar hypothesis discussed in Sect. 4.2 is ruled out, it is still interesting to question our energy reconstruction at low energies. Specifically, if we *did* have a contamination of ^{37}Ar at a level required to explain the excess, how often might we mis-reconstruct the mean of the peak to be 2.3 keV instead of the nominal 2.8 keV? The likelihood ratio curve in Fig. 4.23 (left) suggests that this is expected to be quite rare, at the level of $\gtrsim 2\sigma$; however, this assumes a pure Gaussian spectrum for ^{37}Ar, which we know is not quite true as we approach lower energies. In order to test if this assumption would lead to incorrectly reconstructed peaks, we use ^{37}Ar calibration data.

To test whether we might reconstruct the mean of an ^{37}Ar peak at 2.3 keV, we simulated many background datasets with ^{37}Ar injected, where the true distribution of ^{37}Ar from calibration data (Fig. 4.6) was used. For the amount of argon injected, we assumed a Poisson distribution with mean rate of 65 events/(t·y), given by the best-fit value from fitting a 2.8 keV peak to the SR1 data. These toy datasets were then fitted with a B_0 + peak hypothesis, where the mean of the peak was unconstrained between 1.7 and 3.3 keV, just as above. In this way, the simulated datasets are data-driven, while the interpretation of those datasets is identical to that shown in Fig. 4.23. The resulting distribution of peak means is shown in Fig. 4.24, where it can be seen that the distribution is centered near the true ^{37}Ar value of 2.82 keV. A reconstructed peak mean of 2.3 keV is quite rare in this study, at the level 2.4σ. Note that this is a larger deviation than that given by the likelihood ratio curve in Fig. 4.23, presumably due to the fact the ^{37}Ar calibration spectrum has a better resolution than that expected from our model. This data-driven study validates Fig. 4.23; if the excess were indeed from a monoenergetic source, our analysis suggests 2.3 ± 0.2 keV line.

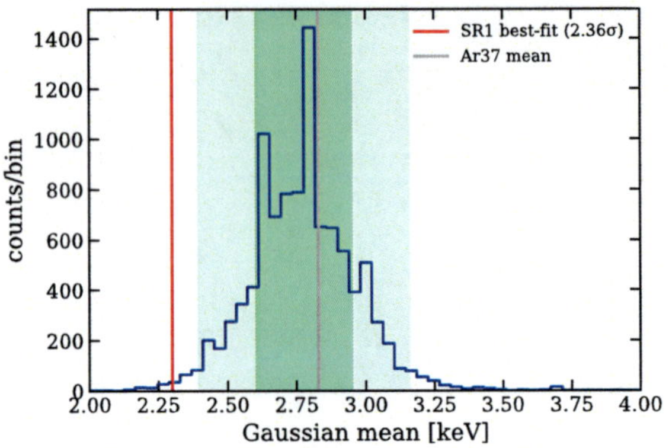

Fig. 4.24 Result of toy-MC study where ^{37}Ar events drawn from the observed distribution in calibration data and injected into simulated background datasets, and then fit with an unconstrained Gaussian model. The distribution of fitted peak means is centered around the expected value of ~2.8 keV, and a 2.3 keV reconstructed peak is disfavored at 2.4σ

4.8 Science Run 2

In an attempt to further study the result we observed in SR1, we also checked data from Science Run 2 (SR2), an R&D science run that followed SR1. Many purification upgrades were implemented during SR2, including the replacement of the xenon circulation pumps with units that (1) are more powerful, leading to improved purification speed, and (2) have lower ^{222}Rn emanation, leading to a reduced ^{214}Pb background rate in the TPC [38, 64], which is further decreased by online radon distillation. The resulting increased purification speed and reduced background make SR2 useful to study the tritium hypothesis. If the excess were from tritium (or another non-noble contaminant), we would expect its rate to decrease due to the improved purification; on the other hand, the rate of the signal hypotheses would not change with purification speed.

While the SR2 purification upgrades allowed for an improved xenon purity and a reduced background level, the unavoidable interruption of recirculation for the upgrades also led to less stable detector conditions. Thus, in addition to a similar event selection process as SR1 in Sect. 2.3, we removed several periods of SR2 for this analysis to ensure data quality. Periods where the electron lifetime changed rapidly due to tests of the purification system were removed to reduce uncertainty in the energy reconstruction. We also removed datasets during which a ^{83m}Kr source was left open for calibration. Data within 50 days of the end of neutron calibrations were also removed to reduce neutron-activated backgrounds and better constrain the background at low energies. After the other selections, this data would have only added ~10 days of live time; thus, for simplicity, it was removed rather than fit separately like the SR1 dataset. With these selections, the effective SR2 live time for this analysis is 24.4 days, with an average ER background reduction of 20% in (1, 30) keV as compared to SR1.

A profile likelihood analysis was then performed on SR2 with a similar background model as SR1, denoted as B_{SR2}. This background model was built using the time dependencies of each component, as described in Sect. 3.1. The ^{214}Pb

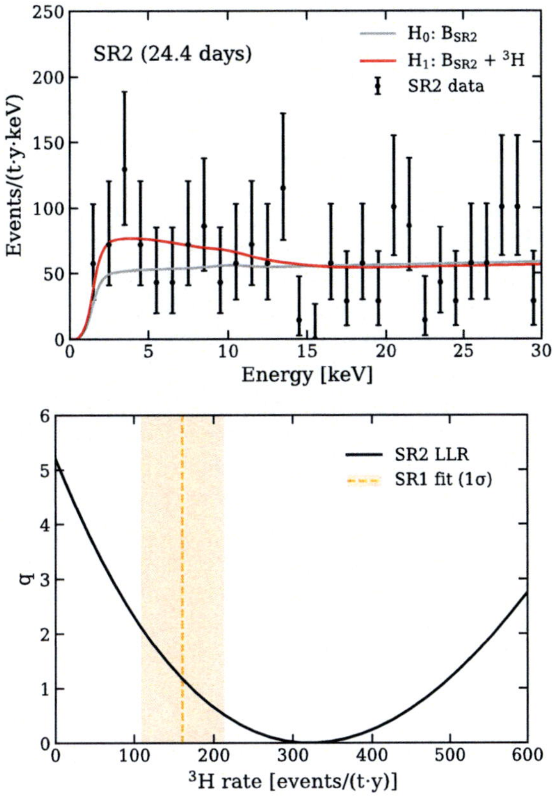

Fig. 4.25 Top: fit of SR2 data to the $B_{SR2} +^3$ H (red) and B_{SR2}-only hypotheses, where a small excess is again observed, albeit with few statistics. Bottom: likelihood ratio curve of the tritium rate, showing that the SR2 best-fit is higher than SR1, but consistent within uncertainties

component was left unconstrained due to the large changes in rate in this time period. Since we are primarily interested in using this data set to test the tritium hypothesis, we focus on the tritium results.

Similarly to SR1, we search for a tritium signal on top of the background model B_{SR2}, and find that the background-only hypothesis is slightly disfavored at $2.3\,\sigma$. The SR2 spectrum, along with the fits for the null (B_{SR2}) and alternative ($B_{\text{SR2}}+^3\text{H}$) hypotheses, can be found in Fig. 4.25 (top). A log-likelihood ratio curve for the tritium component is given in Fig. 4.25 (bottom), which shows that the fitted tritium rate is 320 ± 160 events/(t·y), higher than that from SR1 but consistent within uncertainties. The rate uncertainty in SR2 is much larger than that in SR1 due to limited statistics. The solar axion and magnetic moment hypotheses give similar results, with significances $\sim 2\,\sigma$ and best-fit values larger than, but consistent with, the respective SR1 fit results. Thus these SR2 studies are largely inconclusive.

4.9 Time Dependence

It is useful to consider the time dependence of the XENON1T excess in the context of the hypotheses discussed up to this point. Due to the changing detector conditions in SR2, we conservatively focus only on the time dependence in SR1, which is much more stable.

Under the tritium hypothesis, we would expect the excess to decay with a half life of 12.3 years; we do not have the ability to observe this decay with a live time of <1 year. The solar axion and neutrino magnetic moment signals would display a $\sim 7\%$ modulation (peak to peak) due to the changing Earth-Sun distance throughout the year, which again is a subtle effect that is unlikely to be observed with the statistics we have. Bosonic dark matter would also display an annual modulation, with a larger amplitude than the solar signals and with a different phase.

At this time, we cannot exclude any hypothesis (except for the background-only one) on the basis of time dependence. In Fig. 4.26 (top) we show the event rate between 1 and 7 keV compared to the expectation from integrating the B_0-only fit. A binning of 25 days was used, and the bin center is given by the livetime-weighted

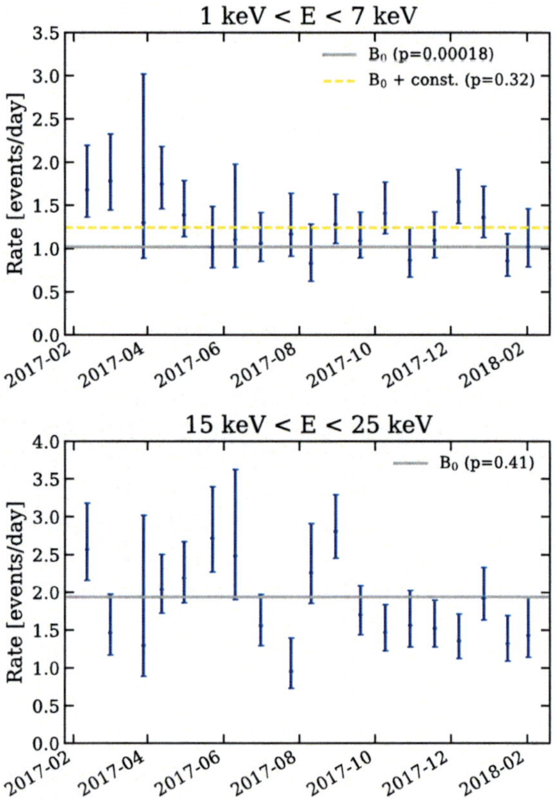

Fig. 4.26 Top: time dependence of the event rate between 1 and 7 keV during SR1, along with the expectation from the B_0 best fit and a fit to the event rate of a simple constant. Bottom: event rate in a side band of 15–25 keV, where good agreement with B_0 is observed

mean in each bin. Due to low numbers of counts, asymmetric Poisson errors are assumed for the data points, and a p-value of $p = 0.0002$ was calculated using Fisher's method, implying that the B_0 model is rejected at 3.5σ on the basis of time dependence. The same procedure was done on a side band region of 15–25 keV, where good agreement is observed between B_0 and the data ($p = 0.4$), as shown in Fig. 4.26 (bottom).

By eye it appears that the excess rate might be elevated in the early parts of SR1. However, a fit with a B_0 + const. model, where the *const.* is accounting for the excess rate and assuming it is constant in time, the p-value improves to $p = 0.3$, meaning that we cannot exclude a constant-in-time hypothesis for the excess. This fit is also included in Fig. 4.26 (top) for comparison. All other hypotheses considered, namely solar axions, neutrino magnetic moment, and tritium, are also consistent with the data, with similar p-values. A more detailed time dependence study or—even better—a more sensitive detector with longer exposure is needed in order to further investigate the temporal behavior of the observed excess. We revisit this question briefly in Sect. 4.10.

4.10 Updated Analysis

Here we consider a number of updates to the analysis, which were considered after publication of the main results. We briefly highlight the various updates and then discuss the results. Since these results rely on work that has not been peer reviewed, they should be considered supplementary to, but not replacements for, the main analysis described in the preceding sections.

4.10.1 Summary of Updates

Combination of SR1 and SR2 For this updated analysis we combine SR1 and SR2, meaning that we fit simultaneously three datasets: $\mathrm{SR1}_a$, $\mathrm{SR1}_b$, and SR2. We use the time dependence of the different background components as described in Sect. 3.1. The ^{214}Pb model is unconstrained and shared between the two SR1 partitions, and its rate in SR2 is constrained to be within 5% of the SR1 value, based on the uncertainty from the time evolution of the ^{218}Po alphas. As discussed below, we also include the time dependence of the solar ν background, which is not expected to change the results but is more consistent with how the other backgrounds are treated. This leaves the materials component, ^{136}Xe, and ^{124}Xe as the only constant backgrounds across all three datasets.

Update to the ^{214}Pb Model There has been recent work on the energy spectrum of the β-decay of ^{214}Pb [109], where it was for the first time calculated as a forbidden transition including the atomic screening and exchange effects. This calculation

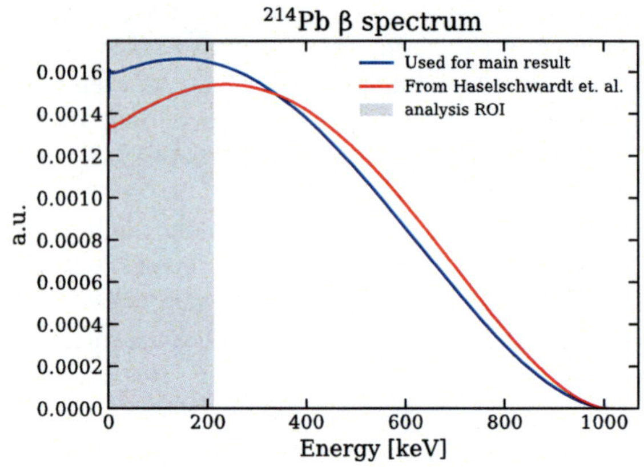

Fig. 4.27 The recently calculated ^{214}Pb model (red) from [109], compared to the model used for the main result for this work (navy). The analysis region of interest is also shown

predicts a ∼15–20% lower event rate from ^{214}Pb in the energy region of 1–15 keV, and is thus quite pertinent to this analysis. For the updated analysis described in this section, we use this calculation of the spectrum with a weak axial vector coupling of $g_{aA} = 0.85$, the median value considered by the authors in [109]. Figure 4.27 shows a comparison of the new model with the one used for the main result of this work. The overall rate is unconstrained for SR1, as in the main analysis. The result of the SR1 fit is used as an input to the SR2 model, along with the time dependence in Fig. 3.1 (right), to give a predicted SR2 rate. An uncertainty of 5% is included for the SR2 rate, based on the time dependence of the ^{218}Po alphas.

Update to the ^{136}Xe Model In the main analysis we included a 10% uncertainty on the ^{136}Xe rate, due in part to the theoretical uncertainty on the spectrum from not knowing which model of 2νββ-decay (HSD or SSD) is realized for this nucleus. After release of the main results, we were given a calculation of the SSD model (Jenni Kotila, private communication), allowing for a more well-motivated background model for this component. The differences between the two spectra, shown in Fig. 4.28, are minor, with the SSD model displaying a slightly lower event rate at the lowest energies. For the updated analysis, we consider the difference between the two models more directly by introducing an additional shape parameter. The updated ^{136}Xe model M is given by

$$M = (1 - f)M_{\text{HSD}} + f M_{\text{SSD}},$$

where M_{HSD} is the HSD model (used for the main analysis), M_{SSD} is the SSD one, and f is the new shape parameter, defined between 0 and 1, which is used to let the data choose which model (or a mix of the two) it prefers. Treating the spectral uncertainty in this way, we then use a tighter uncertainty on the ^{136} Xe rate of 3.2%, based on the measured abundance and the half-life [15].

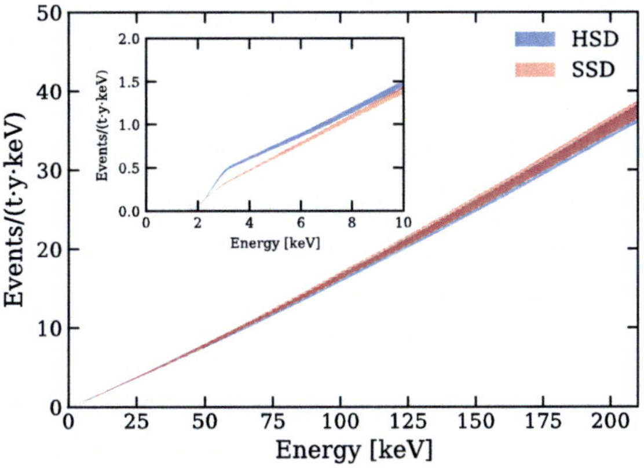

Fig. 4.28 Comparison of the ^{136}Xe model under the higher-state dominates (HSD) model and the single-state dominates (SSD) model. The HSD model was used for the main result and given in [120, 121]

Time Dependence of Signal Models In light of the discussion above on time dependence, for the updated analysis we include time dependence in the signal models, which was omitted for the main inference. For the solar axion and magnetic moment hypotheses, as well as the solar ν background, we consider the changing flux due to the Earth-Sun distance throughout the year. The bosonic dark matter hypothesis includes the expected annual modulation from the relative direction of Earth's velocity to the Sun's. Lastly, the tritium hypothesis includes its decay with a $T_{1/2} = 12.3$ years. For all of these models, no uncertainty on the time dependence is included; a fixed correction is simply applied to the rate based on the time of each event in the three datasets. For the solar axion and neutrino magnetic moment these factors are given by 0.98, 1.01, and 1.00 for $\mathrm{SR1_a}$, $\mathrm{SR1_b}$, and SR2, respectively, and for tritium[10] they are 0.98, 0.96, and 0.93.

4.10.2 Results of Updated Analysis

First we do a background-only fit with the new background model, denoted $\hat{B}_0$. The results are shown in Fig. 4.29, where the SR1 and SR2 fits are both included, along with the sum of the two. Note that the SR1 fit is still done for the two partitions $\mathrm{SR1}_a$ and $\mathrm{SR1}_b$, but we omit their individual plots for brevity. Compared to the main background model B_0, the updated $\hat{B}_0$ fit gives a lower rate at low energies due to the change of the ^{214}Pb model.

To simplify the inference of this updated analysis, we consider an ABC-only solar axion model, along with the neutrino magnetic moment and tritium

[10] Based on an arbitrarily chosen value of t_0 of February 1, 2017.

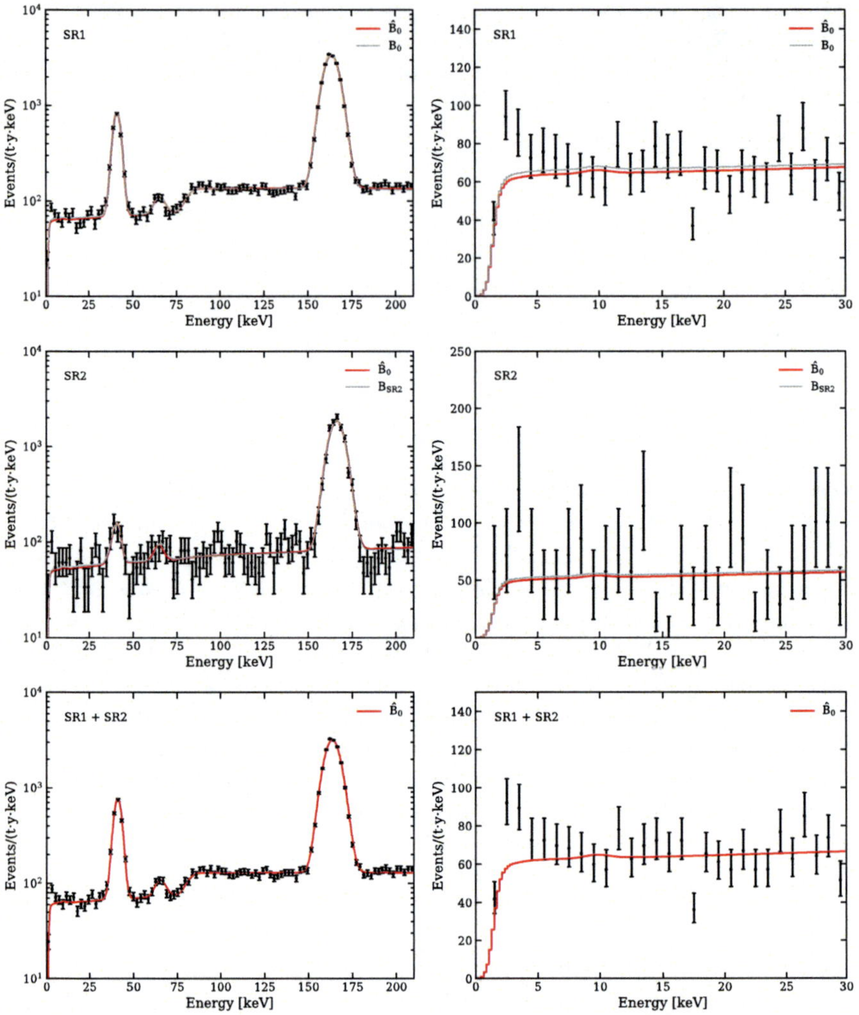

Fig. 4.29 Summary of the updated analysis, which fits SR1 and SR2 simultaneously and includes updates to the background/signal models. The left column shows the full analysis ROI and the right column the zoom to low energy; and the rows show the SR1, SR2, and SR1+SR2 fit results, respectively

hypotheses. This way we can safely assume asymptotic behavior of the test statistic and not rely on toy-MC methods, which require significant computation time.

The fits including signal are summarized in Fig. 4.30. After implementing all the updates described above, the observed excess displays a larger significance, primarily due to (1) the combination of SR1 and SR2 and (2) the updated ^{214}Pb model. Under the solar axion signal hypothesis, the $\hat{B}_0$ is rejected at 4.7σ; compared

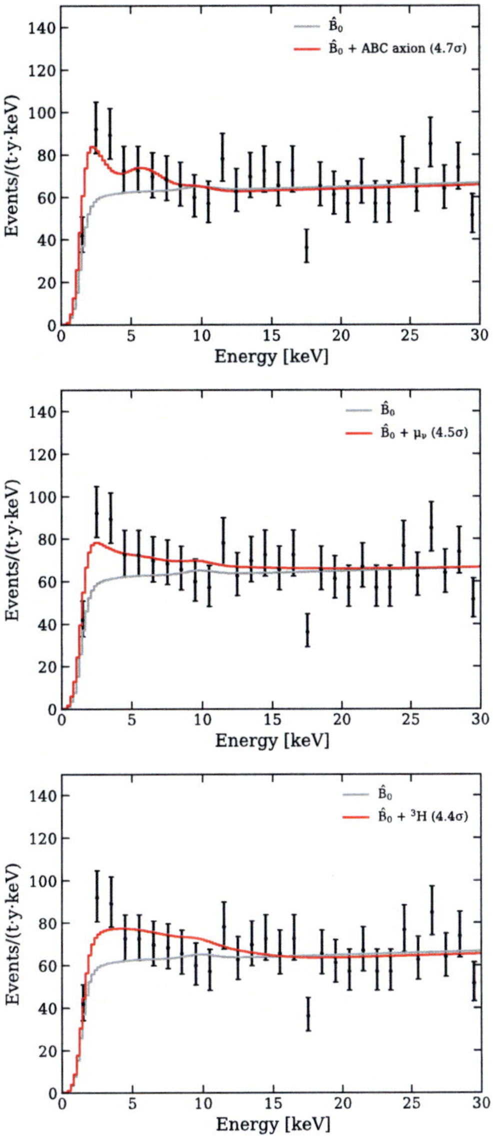

Fig. 4.30 Fits including the solar axion (top), magnetic moment (middle), and tritium (bottom) hypotheses compared to the updated background model $\hat{B}_0$

to the magnetic moment hypothesis $\hat{B}_0$ is rejected at 4.5σ; and for the tritium hypothesis it is 4.4σ. If we removed SR2 from the analysis, the significances would be 4.1σ, 3.9σ, and 3.8σ, respectively. Thus, while this updated analysis does give stronger evidence for the present of an excess, it does not help distinguish between the different hypotheses outlined in the previous sections.

Chapter 5
Conclusions and Outlook

Abstract This chapter summarizes the work presented in this thesis and comments on the future outlook of the XENON program: the XENONnT experiment.

5.1 Summary

This thesis has described the results of a search for new physics in XENON1T in the electronic recoil channel. Chapter 1 presented the XENON1T detector and reviewed the theoretical motivation for solar axions, a neutrino magnetic moment, and bosonic dark matter. Chapter 2 detailed the energy reconstruction and energy threshold of XENON1T. Chapter 3 characterized the background components in terms of both their energy spectra and time dependence, and described the statistical framework for the three different signal models. Lastly, Chap. 4 reported the results of this search for new physics.

We observe an excess at low energies (1–7 keV), where 285 events are observed compared to an expected 232 events from the background-only fit to the data. There are three possible explanations for this excess: a statistical flucttutation, a previously-unconsidered background component, or a hint of new physics.

Statistical Fluctuation With a statistical significance of $\gtrsim 3\sigma$, the SR1 result would be expected under the background-only hypothesis in $\lesssim 0.1\%$ of experiments. While significant, this does not completely rule out the possibility of statistical fluctuation. However, the updated analysis presented in Sect. 4.10 gives an even larger significance ($\gtrsim 4.5\sigma$) due to the combination of a larger exposure and lower predicted background rate at low energies. This discounts the statistical fluctuation hypothesis.

New Background The β decay of tritium is considered a possible explanation for the excess, as it has a similar spectrum to that observed and is expected to be present in the detector at some level. We are unable to independently confirm the presence of tritium at the $O(10^{-25})$ mol/mol concentration required to account for the excess, and so treat it separately from our validated background model. If electronic recoils

E. Shockley, *Study of Excess Electronic Recoil Events in XENON1T*,
Springer Theses, https://doi.org/10.1007/978-3-030-87752-1_5

from tritium decay were the source of the excess, this would be its first indication as an atmospheric source of background in LXe TPCs.

New Physics The tritium hypothesis clearly represents a possible SM explanation for the excess, but—based on spectral shape alone—the solar axion model is the most favored by the data at 3.4σ, albeit at only $\sim 2\sigma$ if one considers tritium as an additional background.

Were it a hint of a solar axion signal, our result would suggest either (1) a non-zero rate of ABC axions or (2) a non-zero rate of both Primakoff and ^{57}Fe axions. If we interpret the excess as an ABC axion signal (i.e., take $g_{a\gamma}$ and g_{an}^{eff} to be zero), the required value of g_{ae} is smaller than that ruled out by other direct searches but has a clear discrepancy with constraints from indirect searches [72, 90]. These constraints are a factor of $\sim$5–10 lower than reported here, although subject to systematic uncertainties. It is noteworthy that some of these astrophysical analyses, while their constraints are still stronger than direct searches, do in fact suggest an additional source of cooling compatible with axions [72, 101]. If the indirect hints and the XENON1T excess were indeed explained by axions, the tension in g_{ae} could be relieved by underestimated systematic uncertainties in, e.g., stellar evolution theory [90] or white dwarf luminosity functions [59], an additional detection mechanism [78, 100], or by a larger solar axion flux than that considered here (see Sect. 1.2.1). In Sect. 4.5.2, we interpreted the excess in the context of benchmark QCD axion models, where a DFSZ-type axion with $m_a \sim 0.1$–4.2 eV could explain the excess; such an axion is not ruled out by any direct detection constraints, but as discussed would be in tension with astrophysical constraints. A KSVZ-type axion does not described the excess well.

Additionally, we describe a direct search for an enhanced neutrino magnetic moment. This signal also has a similar spectrum to the excess observed, but at $3.2\,\sigma$ displays a lower significance than that from solar axions. We report a confidence interval of

$$\mu_\nu \in [1.4,\ 2.9] \times 10^{-11}\,\mu_B\ (90\%\ C.L.),$$

the upper boundary of which is very close to the world-leading direct limit reported by Borexino [7]. This shows that dark matter experiments are also sensitive to beyond-SM physics in the neutrino sector. Here we only search for an enhanced neutrino-electron cross-section due to an anomalous magnetic moment, but a similar enhancement would also occur in neutrino-nucleus scattering [107]. With the discrimination capabilities of LXe TPCs to ER and NR events, it would be interesting to consider this channel in future searches as well.

We also search for bosonic dark matter, which would express itself as a monoenergetic peak at the rest mass of the particle. Our search originally just considered masses in 1-keV steps, where no excess above 3σ was observed, but after observation of the excess we generalized the search at low energies. A 2.3 keV peak describes well the excess we observe, with a significance $\sim 3\sigma$ after accounting for the look elsewhere effect. While this peak is noticeably close to the expected peak from an ^{37}Ar background, we can exclude this hypothesis based on measurements of the argon concentration in the lab air outside the detector, and we include studies with calibration that show the observed excess does not really look like ^{37}Ar.

Finally, we include an update to the analysis. This update includes some preliminary results that have not yet been published, and so should be considered supplemental material to the main result and not a replacement for it. The update, due primarily to a lower predicted rate from the dominant background ^{214}Pb and the combination of SR1 and SR2, gives an even larger significance for the excess, 4.7σ and 4.5σ under the solar axion and magnetic moment hypotheses, respectively. However, with tritium hypothesis is still viable, so these significances cannot be used to claim a discovery of new physics. It is unlikely that we will understand the nature of the excess with the current data from XENON1T, and so a larger, more sensitive detector is likely needed. Such a detector, XENONnT, is being currently being commissioned.

5.2 Outlook: XENONnT

The next phase of the XENON program, XENONnT, is currently being commissioned, with data expected in 2021. Featuring a 5.9t active volume and a projected reduction in background by a factor of 6 [41], XENONnT will reach further unprecedented sensitivity to rare-event signals, including the ones discussed in this work.

If the excess observed in XENON1T persists in the XENONnT upgrade, we will be able to further study it and possibly identify its origin. Is it a new background, a sign of new physics, or a statistical fluke? The most interesting case would of course be a sign of new physics, which we briefly study here.

We use the framework presented in this work to estimate the sensitivity of XENONnT. Assuming a 4t fiducial volume and the same efficiencies as XENON1T, we simulate many datasets with an ABC solar axion signal injected, with a g_{ae} given by the XENON1T best fit. For each of these datasets, we compare hypotheses H_0 : $B_0^{\mathrm{nt}} +^3$ H and H_1 : $B_0^{\mathrm{nt}} +^3$ Haxion, where B_0^{nt} is the XENONnT background model and with the tritium component unconstrained in both cases. This test allows us to determine the expected significance at which we can reject the tritium explanation for the excess. We performed this procedure for different concentrations of ^{214}Pb, and as a function of total exposure of XENONnT. The "discrimination significance", the expected significance at which we can exclude tritium assuming a solar axion signal, is shown in Fig. 5.1 as a function of livetime and for different background levels, where one can see that XENONnT will be able to differentiate these two spectra at the $\sim 5\sigma$ level within a few months of data-taking.

XENONnT is not the only experiment that will be able to confirm or exclude the XENON1T result. The LZ experiment [13] and PandaX-4t [166] will also be

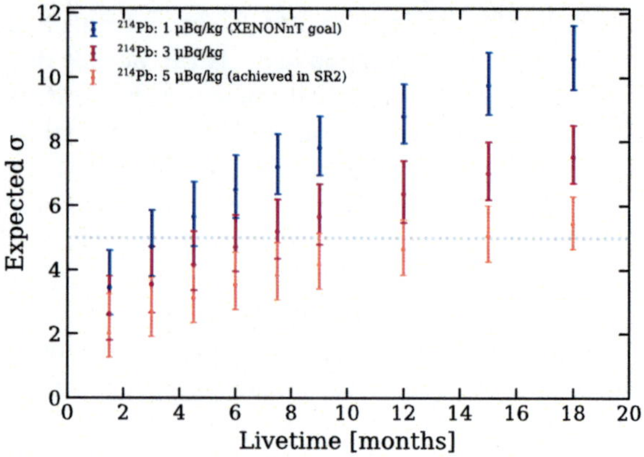

Fig. 5.1 The expected significance at which XENONnT could exclude the tritium hypothesis if the XENON1T excess were due to a solar axion signal. Shown as a function of livetime and for different concentrations of ^{214}Pb, the dominant background at low energies

coming online soon, with similar schedules and similar sensitivities. It will be very interesting to see what these three experiment find.

5.3 Conclusion

Liquid xenon time projection chambers like XENON1T were originally designed to search for WIMP dark matter. Moreover, as low-background, low-energy threshold, and large-exposure particle detectors, they are sensitive to a multitude of other physics beyond the SM. In this thesis we considered a search for solar axions, neutrino magnetic moment, and bosonic dark matter using the XENON1T detector. Interestingly, we observe an excess that could be consistent with some of these signals. This motivates a broader focus of liquid xenon TPCs and increase the priority of other interesting signals to the same level as the WIMP search.

The XENONnT experiment will investigate the XENON1T result with an increased sensitivity. It will be very interesting to see the results in XENONnT and other similar experiments, and we fortunately do not have to wait long.

References

1. J. Aalbers, Dark matter search with XENON1T. PhD thesis, Amsterdam University, 2018
2. J. Aalbers, C. Tunnell, PAX: processor for analyzing XENON, Version 6.10.1. https://github.com/XENON1T/pax
3. L.F. Abbott, P. Sikivie, A cosmological bound on the invisible axion. Phys. Lett. B **120**(1), 133–136 (1983)
4. K. Abe et al., Search for dark matter in the form of hidden photons and axion-like particles in the XMASS detector. Phys. Lett. B **787**, 153–158 (2018)
5. N. Abgrall et al., New limits on bosonic dark matter, solar axions, Pauli exclusion principle violation, and electron decay from the majorana demonstrator. Phys. Rev. Lett. **118**, 161801 (2017)
6. S. Agostinelli et al., GEANT4–a simulation toolkit, SLAC-PUB-9350, FERMILAB-PUB-03-339, CERN-IT-2002-003. Nucl. Instrum. Meth. A. **506**, 250–303 (2003)
7. M. Agostini et al., Limiting neutrino magnetic moments with Borexino Phase-II solar neutrino data. Phys. Rev. D **96**, 091103 (2017)
8. M. Agostini et al., The first search for bosonic super-WIMPs with masses up to 1 MeV/c^2 with GERDA, May 2020. arXiv:2005.14184
9. D.S. Akerib et al., Tritium calibration of the LUX dark matter experiment. Phys. Rev. D **93**, 072009 (2016)
10. D.S. Akerib et al., Improved limits on scattering of weakly interacting massive particles from reanalysis of 2013 LUX data. Phys. Rev. Lett. **116**(16), 161301 (2016). arXiv:1512.03506
11. D.S. Akerib et al., First searches for axions and axionlike particles with the LUX experiment. Phys. Rev. Lett. **118**, 261301 (2017)
12. D.S. Akerib et al., Search for annual and diurnal rate modulations in the LUX experiment. Phys. Rev. D **98**(6), 062005 (2018). arXiv:1807.07113
13. D.S. Akerib et al., The LUX-ZEPLIN (LZ) experiment. Nucl. Instrum. Methods Phys. Res. Sect. A **953**, 163047 (2020)
14. Y. Akrami et al., Planck 2018 results. I. Overview and the cosmological legacy of Planck, July 2018. arXiv:1807.06205
15. J.B. Albert et al., Improved measurement of the $2\nu\beta\beta$ half-life of ^{136}Xe with the EXO-200 detector. Phys. Rev. C **89**, 015502 (2014)
16. F. Alessandria et al., Search for 14.4 keV solar axions from M1 transition of57Fe with CUORE crystals. J. Cosmol. Astropart. Phys. **2013**(05), 007 (2013)
17. S. Algeri, J. Aalbers, K.D. Morå, J. Conrad, Searching for new phenomena with profile likelihood ratio tests. Nat. Rev. Phys. **2**(5), 245–252 (2020)

E. Shockley, *Study of Excess Electronic Recoil Events in XENON1T*,
Springer Theses, https://doi.org/10.1007/978-3-030-87752-1

18. J. Amare et al., Cosmogenic production of tritium in dark matter detectors. Astropart. Phys. **97**, 96–105 (2018). arXiv:1706.05818
19. H. An, M. Pospelov, J. Pradler, A. Ritz, Direct detection constraints on dark photon dark matter. Phys. Lett. B **747**, 331–338 (2015)
20. S Andriamonje et al., Search for 14.4 keV solar axions emitted in the M1-transition of 57Fe nuclei with CAST. J. Cosmol. Astropart. Phys. **2009**(12), 002 (2009)
21. G. Anton et al. Measurement of the scintillation and ionization response of liquid xenon at MeV energies in the EXO-200 experiment. Phys. Rev. C **101**(6), 065501 (2020). arXiv:1908.04128
22. N. Aoki, Y. Makide, The concentration of krypton in the atmosphere—its revision after half a century—. Chem. Lett. **34**, 1396–1397 (2005)
23. E. Aprile, T. Doke, Liquid xenon detectors for particle physics and astrophysics. Rev. Mod. Phys. **82**, 2053–2097 (2010)
24. E. Aprile, R. Mukherjee, M. Suzuki, Measurements of the lifetime of conduction electrons in liquid xenon. Nucl. Instrum. Meth. A **300**, 343–350 (1991)
25. E. Aprile et al., Physics reach of the XENON1T dark matter experiment. J. Cosmol. Astropart. Phys. **2016**(04), 027 (2016)
26. E. Aprile et al., First dark matter search results from the XENON1T experiment. Phys. Rev. Lett. **119**, 181301 (2017)
27. E. Aprile et al., Material radioassay and selection for the XENON1T dark matter experiment. Eur. Phys. J. C **77**(12), 890 (2017)
28. E. Aprile et al., Removing krypton from xenon by cryogenic distillation to the ppq level. Eur. Phys. J. C **77**(5), 275 (2017)
29. E. Aprile et al., Search for bosonic super-WIMP interactions with the XENON100 experiment. Phys. Rev. D **96**, 122002 (2017)
30. E. Aprile et al., The XENON1T dark matter experiment. Eur. Phys. J. C **77**(12), 881 (2017). arXiv:1708.07051
31. E. Aprile et al., Dark matter search results from a one ton-year exposure of XENON1T. Phys. Rev. Lett. **121**, 111302 (2018)
32. E. Aprile et al., Signal yields of keV electronic recoils and their discrimination from nuclear recoils in liquid xenon. Phys. Rev. D **97**, 092007 (2018)
33. E. Aprile et al., Light dark matter search with ionization signals in XENON1T. Phys. Rev. Lett. **123**, 251801 (2019)
34. E. Aprile et al., Observation of two-neutrino double electron capture in 124Xe with XENON1T. Nature **568**(7753), 532–535 (2019)
35. E. Aprile et al., The XENON1T data acquisition system. J. Instrum. **14**(07), P07016 (2019). arXiv:1906.00819
36. E. Aprile et al., XENON1T dark matter data analysis: signal and background models and statistical inference. Phys. Rev. D **99**, 112009 (2019)
37. E. Aprile et al., XENON1T dark matter data analysis: signal reconstruction, calibration, and event selection. Phys. Rev. D **100**, 052014 (2019)
38. E. Aprile et al., ^{222}Rn emanation measurements for the XENON1T experiment, Sept 2020. arXiv:2009.13981
39. E. Aprile et al., Energy resolution and linearity of XENON1T in the MeV energy range. Eur. Phys. J. C **80**(8), 785 (2020). arXiv:2003.03825
40. E. Aprile et al., Excess electronic recoil events in XENON1T, June 2020. arXiv:2006.09721
41. E. Aprile et al., Projected WIMP sensitivity of the XENONnT dark matter experiment, July 2020. arXiv:2007.08796
42. E. Aprile et al., Application and modeling of an online distillation method to reduce Krypton and Argon in XENON1T. In preparation
43. T. Aralis et al., Constraints on dark photons and axion-like particles from SuperCDMS Soudan (2019). arXiv:1911.11905
44. S. Arceo-Díaz, K.P. Schröder, K. Zuber, D. Jack, Constraint on the magnetic dipole moment of neutrinos by the tip-RGB luminosity in ω-Centauri. Astropart. Phys. **70**, 1–11 (2015)

45. P. Arias, D. Cadamuro, M. Goodsell, J. Jaeckel, J. Redondo, A. Ringwald, WISPy cold dark matter. J. Cosmol. Astropart. Phys. **2012**(06), 013 (2012)
46. K. Arisaka, P. Beltrame, C. Ghag, J. Kaidi, K. Lung, A. Lyashenko, R.D. Peccei, P. Smith, K. Ye, Expected sensitivity to galactic/solar axions and bosonic super-WIMPs based on the axio-electric effect in liquid xenon dark matter detectors. Astropart. Phys. **44**, 59–67 (2013)
47. E. Armengaud et al., Searches for electron interactions induced by new physics in the EDELWEISS-III germanium bolometers. Phys. Rev. D **98**, 082004 (2018)
48. A. Ayala, I. Domínguez, M. Giannotti, A. Mirizzi, O. Straniero, Revisiting the bound on axion-photon coupling from globular clusters. Phys. Rev. Lett. **113**, 191302 (2014)
49. O. Azzolini et al., Evidence of single state dominance in the two-neutrino double- β decay of Se 82 with CUPID-0. Phys. Rev. Lett. **123**(26), 262501 (2019)
50. J.N. Bahcall, C. Peña-Garay, Solar models and solar neutrino oscillations. New J. Phys. **6**, 63 (2004)
51. P. Barrow et al., Qualification tests of the R11410-21 photomultiplier tubes for the XENON1T detector. J. Instrum. **12**(01), P01024 (2017). arXiv:1609.01654
52. V.I. Barsanov et al., Artificial neutrino source based on the Ar-37 isotope. Phys. Atom. Nucl. **70**, 300–310 (2007)
53. K Barth et al., CAST constraints on the axion-electron coupling. J. Cosmol. Astropart. Phys. **2013**(05), 010 (2013)
54. M.-M. Bé, V. Chisté, C. Dulieu, X. Mougeot, V. Chechev, N. Kuzmenko, F. Kondev, A. Luca, M. Galán, A.L. Nichols, A. Arinc, A. Pearce, X. Huang, B. Wang, *Table of Radionuclides. Monographie BIPM-5*, vol. 6 (Bureau International des Poids et Mesures, Pavillon de Breteuil, Sèvres, 2011)
55. M.-M. Bé, V. Chisté, C. Dulieu, X. Mougeot, V.P. Chechev, F.G. Kondev, A.L. Nichols, X. Huang, B. Wang, *Table of Radionuclides. Monographie BIPM-5*, vol. 7 (Bureau International des Poids et Mesures, Pavillon de Breteuil, Sèvres, 2013)
56. A.G. Beda, V.B. Brudanin, V.G. Egorov, D.V. Medvedev, V.S. Pogosov, E.A. Shevchik, M.V. Shirchenko, A.S. Starostin, I.V. Zhitnikov, Gemma experiment: the results of neutrino magnetic moment search. Phys. Part. Nucl. Lett. **10**(2), 139–143 (2013)
57. N.F. Bell, V. Cirigliano, M.J. Ramsey-Musolf, P. Vogel, M.B. Wise, How magnetic is the dirac neutrino? Phys. Rev. Lett. **95**(15), 1–4 (2005)
58. N.F. Bell, M. Gorchtein, M.J. Ramsey-Musolf, P. Vogel, P. Wang, Model independent bounds on magnetic moments of Majorana neutrinos. Phys. Lett. Sect. B **642**(4), 377–383 (2006)
59. M.M.M. Bertolami, B.E. Melendez, L.G. Althaus, J. Isern, Revisiting the axion bounds from the Galactic white dwarf luminosity function. J. Cosmol. Astropart. Phys. **2014**(10), 069 (2014)
60. G. Bertone, D. Hooper, J. Silk, Particle dark matter: evidence, candidates and constraints. Phys. Rep. **405**(5), 279–390 (2005)
61. M.V. Beznogov, E. Rrapaj, D. Page, S. Reddy, Constraints on axion-like particles and nucleon pairing in dense matter from the hot neutron star in HESS J1731-347. Phys. Rev. C **98**(3), 035802 (2018). arXiv:1806.07991
62. M.R. Bhat, *Evaluated Nuclear Structure Data File (ENSDF)*. Nucl. Data Sci. Tech. (Springer, Berlin, 1992). Data extracted using the NNDC On-Line Data Service from the ENSDF database, files revised as of Oct 2019
63. I.M. Bloch, R. Essig, K. Tobioka, T. Volansky, T.-T. Yu, Searching for dark absorption with direct detection experiments. J. High Energy Phys. **06**, 087 (2017)
64. E. Brown et al., Magnetically-coupled piston pump for high-purity gas applications. Eur. Phys. J. C **78**(7), 604 (2018)
65. D. Cadamuro, J. Redondo, Cosmological bounds on pseudo Nambu-Goldstone bosons. J. Cosmol. Astropart. Phys. **2012**(02), 032 (2012)
66. P. Carenza, T. Fischer, M. Giannotti, G. Guo, G. Martínez-Pinedo, A. Mirizzi, Improved axion emissivity from a supernova via nucleon-nucleon bremsstrahlung. J. Cosmol. Astropart. Phys. **10**(10), 016 (2019). arXiv:1906.11844 [Erratum: J. Cosmol. Astropart. Phys. **05**, E01 (2020)]

67. S. Chang, K. Choi, Hadronic axion window and the big-bang nucleosynthesis. Phys. Lett. B **316**(1), 51–56 (1993)
68. J.H. Chang, R. Essig, S.D. McDermott, Supernova 1987A constraints on sub-GeV dark sectors, millicharged particles, the QCD axion, and an axion-like particle. J. High Energy Phys. **09**, 051 (2018). arXiv:1803.00993
69. J.-W. Chen, H.-C. Chi, C.-P. Liu, C.-P. Wu, Low-energy electronic recoil in xenon detectors by solar neutrinos. Phys. Lett. B **774**, 656–661 (2017)
70. V. Chepel, H. Araujo, Liquid noble gas detectors for low energy particle physics. J. Instrum. **8**, R04001 (2013). arXiv:1207.2292
71. A.H. Córsico, L.G. Althaus, M.M. Miller Bertolami, S.O. Kepler, E. García-Berro, Constraining the neutrino magnetic dipole moment from white dwarf pulsations. J. Cosmol. Astropart. Phys. **2014**(08), 054 (2014)
72. A. Corsico, L. Althaus, M. Bertolami, S. Kepler, Pulsating white dwarfs: new insights. Astron. Astrophys. Rev. **27**, 12 (2019)
73. G. Cowan, *Statistical Data Analysis* (Clarendon Press, Oxford, 1998)
74. G. Cowan, K. Cranmer, E. Gross, O. Vitells, Asymptotic formulae for likelihood-based tests of new physics. Eur. Phys. J. C **71**(2), 1554 (2011)
75. X. Cui et al., Dark matter results from 54-ton-day exposure of pandaX-II experiment. Phys. Rev. Lett. **119**, 181302 (2017)
76. C.E. Dahl, The physics of background discrimination in liquid xenon, and first results from Xenon10 in the hunt for WIMP dark matter. PhD thesis, Princeton University, 2009
77. P.F. de Salas, K. Malhan, K. Freese, K. Hattori, M. Valluri, On the estimation of the local dark matter density using the rotation curve of the Milky Way. J. Cosmol. Astropart. Phys. **2019**(10), 037 (2019)
78. J.B. Dent, B. Dutta, J.L. Newstead, A. Thompson, Inverse Primakoff scattering as a probe of solar axions at liquid xenon direct detection experiments, June 2020. arXiv:2006.15118
79. G.G. di Cortona, E. Hardy, J.P. Vega, G. Villadoro, The QCD axion, precisely. J. High Energy Phys. **2016**(1), 1–37 (2016)
80. L. Di Luzio, F. Mescia, E. Nardi, Window for preferred axion models. Phys. Rev. D **96**, 075003 (2017)
81. L. Di Luzio, M. Giannotti, E. Nardi, L. Visinelli, The landscape of QCD axion models, May 2020. arXiv:2003.01100
82. D.A. Dicus, E.W. Kolb, V.L. Teplitz, R.V. Wagoner, Astrophysical bounds on the masses of axions and Higgs particles. Phys. Rev. D **18**, 1829–1834 (1978)
83. S. Dimopoulos, G.D. Starkman, B.W. Lynn, Atomic enhancements in the detection of axions. Mod. Phys. Lett. A **1**, 491–500 (1986)
84. S. Dimopoulos, J.A. Frieman, B.W. Lynn, G.D. Starkman, Axiorecombination: a new mechanism for stellar axion production. Phys. Lett. B **179**, 223–227 (1986)
85. S. Dimopoulos, G.D. Starkman, B.W. Lynn, Atomic enhancements in the detection of weakly interacting particles. Phys. Lett. B **168**, 145–150 (1986)
86. M. Dine, W. Fischler, The not so harmless axion. Phys. Lett. B **120**, 137–141 (1983)
87. M. Dine, W. Fischler, M. Srednicki, A simple solution to the strong CP problem with a harmless axion. Phys. Lett. B **104**(3), 199–202 (1981)
88. A. Dobi, D.S. Leonard, C. Hall, L. Kaufman, T. Langford, S. Slutsky, Y.R. Yen, Study of a zirconium getter for purification of xenon gas. Nucl. Instrum. Meth. A **620**, 594–598 (2010)
89. M. Doi, T. Kotani, Neutrino emitting modes of double beta decay. Prog. Theor. Phys. **87**(5), 1207–1231 (1992)
90. S.A. Díaz, K.-P. Schröder, K. Zuber, D. Jack, E.E.B. Barrios, Constraint on the axion-electron coupling constant and the neutrino magnetic dipole moment by using the tip-RGB luminosity of fifty globular clusters (2019). arXiv:1910.10568
91. C.H. Faham, V.M. Gehman, A. Currie, A. Dobi, P. Sorensen, R.J. Gaitskell, Measurements of wavelength-dependent double photoelectron emission from single photons in VUV-sensitive photomultiplier tubes. J. Instrum. **10**(09), P09010 (2015)

92. G.J. Feldman, R.D. Cousins, A unified approach to the classical statistical analysis of small signals. Phys. Rev. D **57**(7), 3873–3889 (1997). arXiv:physics/9711021
93. A. Fieguth, First observation of double electron capture in Xe-124 and detection prospects for underlying nuclear interaction mechanisms in direct dark matter search. PhD thesis, Westfälische Wilhelms-Universität Münster, 2018
94. T. Fischer, S. Chakraborty, M. Giannotti, A. Mirizzi, A. Payez, A. Ringwald, Probing axions with the neutrino signal from the next galactic supernova. Phys. Rev. D **94**(8), 085012 (2016). arXiv:1605.08780
95. C. Fu et al., Limits on axion couplings from the first 80 days of data of the pandaX-II experiment. Phys. Rev. Lett. **119**, 181806 (2017)
96. K. Fujikawa, R.E. Shrock, Magnetic moment of a massive neutrino and neutrino-spin rotation. Phys. Rev. Lett. **45**, 963–966 (1980)
97. H.W. Gäggeler, Radioactivity in the atmosphere. Radiochim. Acta **70–71**(s1), 345–354 (1995)
98. R.J. Gaitskell, Direct detection of dark matter. Annu. Rev. Nucl. Part. Sci. **54**(1), 315–359 (2004). https://doi.org/10.1146/annurev.nucl.54.070103.181244
99. P. Galison, A. Manohar, Two Z's or not two Z's? Phys. Lett. B **136**(4), 279–283 (1984)
100. C. Gao, J. Liu, L.-T. Wang, X.-P. Wang, W. Xue, Y.-M. Zhong, Re-examining the solar axion explanation for the XENON1T excess, June 2020. arXiv:2006.14598
101. M. Giannotti, I.G. Irastorza, J. Redondo, A. Ringwald, K. Saikawa, Stellar recipes for axion hunters. J. Cosmol. Astropart. Phys. **2017**(10), 010 (2017)
102. P. Gondolo, G.G. Raffelt, Solar neutrino limit on axions and keV-mass bosons. Phys. Rev. D **79**, 107301 (2009)
103. E. Gross, O. Vitells, Trial factors for the look elsewhere effect in high energy physics. Eur. Phys. J. C **70**(1), 525–530 (2010). arXiv:1005.1891
104. K. Hamaguchi, N. Nagata, K. Yanagi, J. Zheng, Limit on the axion decay constant from the cooling neutron star in Cassiopeia A. Phys. Rev. D **98**(10), 103015 (2018). arXiv:1806.07151
105. Hamamatsu, *Photomultiplier Tubes: Basics and Principles* (2007). https://www.hamamatsu.com/resources/pdf/etd/PMT_handbook_v3aE.pdf
106. J. Happell et al., A history of atmospheric tritium gas (HT) 1950–2002. Tellus **56**(3), 183–193 (2004)
107. R. Harnik, J. Kopp, P.A.N. Machado, Exploring ν signals in dark matter detectors. J. Cosmol. Astropart. Phys. **2012**(07), 026 (2012)
108. M.R. Harston, N.C. Pyper, Exchange effects in β decays of many-electron atoms. Phys. Rev. A **45**, 6282 (1992)
109. S. Haselschwardt, J. Kostensalo, X. Mougeot, J. Suhonen, Improved calculations of beta decay backgrounds to new physics in liquid xenon detectors (2020), pp. 1–7. arXiv:2007.13686
110. C. Hasterok, Gas purity analytics, calibration studies, and background predictions towards the first results of XENON1T. PhD thesis, Ruprecht-Karls-Universität, Heidelberg, 2017
111. S. Hauf et al., Radioactive decays in Geant4. IEEE Trans. Nucl. Sci. **60**(4), 2966–2983 (2013)
112. W.C. Haxton, R.G. Hamish Robertson, A.M. Serenelli, Solar neutrinos: status and prospects. Annu. Rev. Astron. Astrophys. **51**(1), 21–61 (2013)
113. IAEA/WMO, Global network of isotopes in precipitation. The GNIP database (2015). https://nucleus.iaea.org/wiser
114. T. Ishida, Isotope effect and isotope separation: a chemist's view. J. Nucl. Sci. Technol. **39**(4), 407–412 (2002)
115. D.B. Kaplan, Opening the axion window. Nucl. Phys. B **260**(1), 215–226 (1985)
116. Yu. Khazov, I. Mitropolsky, A. Rodionov, Nuclear data sheets for A = 131. Nucl. Data Sheets **107**(11), 2715–2930 (2006)
117. Yu. Khazov, A. Rodionov, F.G. Kondev, Nuclear data sheets for A = 133. Nucl. Data Sheets **112**(4), 855–1113 (2011)
118. J.E. Kim, Weak-interaction singlet and strong CP invariance. Phys. Rev. Lett. **43**, 103–107 (1979)

119. E.W. Kolb, M.S. Turner, *The Early Universe*, vol. 69 (Avalon Publishing, New York, 1990)
120. J. Kotila, F. Iachello, Phase-space factors for double-β decay. Phys. Rev. C **85**, 034316 (2012)
121. J. Kotila et al., Double beta decay phase-space factors. http://nucleartheory.yale.edu. Accessed 1 Oct 2020
122. M. Kuster, G. Raffelt, B. Beltran, *Axions* (Springer, Berlin, 2008)
123. L.B. Leinson, Axion mass limit from observations of the neutron star in Cassiopeia A. J. Cosmol. Astropart. Phys. **08**, 031 (2014). arXiv:1405.6873
124. S. Lindemann, H. Simgen, Krypton assay in xenon at the ppq level using a gas chromatographic system and mass spectrometer. Eur. Phys. J. C **74**(2), 2746 (2014)
125. L.L. Lucas, M.P. Unterweger, Comprehensive review and critical evaluation of the half-life of tritium. J. Res. Natl. Inst. Stand. Technol. **105**(4), 541–549 (2000)
126. A. Manalaysay, T. Marrodán Undagoitia, A. Askin, L. Baudis, A. Behrens, A.D. Ferella, A. Kish, O. Lebeda, R. Santorelli, D. Vénos, A. Vollhardt, Spatially uniform calibration of a liquid xenon detector at low energies using Kr83m. Rev. Sci. Instrum. **81**(7), 073303 (2010)
127. J. Mishima, C.M. Steele, Oxidation of tritium gas under accident and transport conditions. Report LA-UR-02-3803, U.S.. Department of Energy Los Alamos, NM, 2002
128. K.D. Morå, Harmonizing discovery thresholds and reporting two-sided confidence intervals: a modified Feldman & Cousins method. J. Instrum. **14**(02), P02003 (2019)
129. S. Moriyama, Proposal to search for a monochromatic component of solar axions using ^{57}Fe. Phys. Rev. Lett. **75**, 3222–3225 (1995)
130. X. Mougeot, Reliability of usual assumptions in the calculation of β and ν spectra. Phys. Rev. C **91**, 055504 (2015)
131. M. Murra, Intrinsic background reduction by cryogenic distillation for the XENON1T dark matter experiment. PhD thesis, Westfälische Wilhelms-Universität Münster, 2019
132. D.A. Nesterenko, K. Blaum, M. Block, C. Droese, S. Eliseev, F. Herfurth, E. Minaya Ramirez, Yu.N. Novikov, L. Schweikhard, V.M. Shabaev, M.V. Smirnov, I.I. Tupitsyn, K. Zuber, N.A. Zubova, Double-β transformations in isobaric triplets with mass numbers $\mathbf{A} = \mathbf{124}$, 130, and 136. Phys. Rev. C **86**, 044313 (2012)
133. Particle Data Group, Review of particle physics. Prog. Theor. Exp. Phys. **2020**(8), 08, 083C01 (2020). https://academic.oup.com/ptep/article-pdf/2020/8/083C01/33653179/ptaa104.pdf
134. R.D. Peccei, H.R. Quinn, CP conservation in the presence of pseudoparticles. Phys. Rev. Lett. **38**, 1440–1443 (1977)
135. W. Plastino et al., Tritium in water electrolytic enrichment and liquid scintillation counting. Radiat. Meas. **42**, 68–73 (2007)
136. W. Plastino et al., Uranium groundwater anomalies and active normal faulting. J. Radioanal. Nucl. Chem. **288**, 101 (2011)
137. M. Pospelov, A. Ritz, M. Voloshin, Bosonic super-WIMPs as keV-scale dark matter. Phys. Rev. D **78**, 115012 (2008)
138. J. Preskill, M.B. Wise, F. Wilczek, Cosmology of the invisible axion. Phys. Lett. B **120**(1), 127–132 (1983)
139. H. Primakoff, Photo-production of neutral mesons in nuclear electric fields and the mean life of the neutral meson. Phys. Rev. **81**, 899 (1951)
140. G.G. Raffelt, *Astrophysical Axion Bounds* (Springer, Berlin, 2008)
141. J. Redondo, Solar axion flux from the axion-electron coupling. J. Cosmol. Astropart. Phys. **2013**(12), 008 (2013)
142. R. Riedmann, Separation of argon from atmospheric air and measurements of ^{37}Ar for CTBT purposes. PhD thesis, University of Bern, 2011
143. R.A. Riedmann, R. Purtschert, Natural 37Ar concentrations in soil air: implications for monitoring underground nuclear explosions. Environ. Sci. Technol. **45**(20), 8656–8664 (2011)
144. R.A. Riedmann, R. Purtschert, Separation of argon from environmental samples for ar-37 and ar-39 analyses. Sep. Purif. Technol. **170**, 217–223 (2016)
145. R. Saldanha, L. Grandi, Y. Guardincerri, T. Wester, Model independent approach to the single photoelectron calibration of photomultiplier tubes. Nucl. Instrum. Meth. A **863**, 35–46 (2017). arXiv:1602.03150

146. R. Saldanha, H.O. Back, R.H.M. Tsang, T. Alexander, S.R. Elliott, S. Ferrara, E. Mace, C. Overman, M. Zalavadia, Cosmogenic production of ^{39}Ar and ^{37}Ar in argon. Phys. Rev. C **100**, 024608 (2019)
147. M. Schumann, Direct detection of WIMP dark matter: concepts and status. J. Phys. G Nucl. Part Phys. **46**(10), 103003 (2019)
148. A. Sedrakian, Axion cooling of neutron stars. Phys. Rev. D **93**(6), 065044 (2016). arXiv:1512.07828
149. M.A. Shifman, A.I. Vainshtein, V.I. Zakharov. Can confinement ensure natural CP invariance of strong interactions? Nucl. Phys. B **166**(3), 493–506 (1980)
150. M. Srednicki, Axion couplings to matter: (I). CP-conserving parts. Nucl. Phys. B **260**(3), 689–700 (1985)
151. M. Szydagis, N. Barry, K. Kazkaz, J. Mock, D. Stolp, M. Sweany, M. Tripathi, S. Uvarov, N. Walsh, M. Woods, NEST: a comprehensive model for scintillation yield in liquid xenon. J. Instrum. **6**(10), P10002 (2011)
152. M. Tanabashi et al., Review of particle physics. Phys. Rev. **D98**(3), 030001 (2018)
153. K. van Bibber, P.M. McIntyre, D.E. Morris, G.G. Raffelt, Design for a practical laboratory detector for solar axions. Phys. Rev. D **39**, 2089–2099 (1989)
154. M. Verpelli, L. Vrapcenjak, LiveChart of nuclides. IAEA, Nuclear Data Section (2020). https://www-nds.iaea.org/livechart/
155. N. Viaux, M. Catelan, P.B. Stetson, G.G. Raffelt, J. Redondo, A.A.R. Valcarce, A. Weiss, Neutrino and axion bounds from the globular cluster M5 (NGC 5904). Phys. Rev. Lett. **111**, 231301 (2013)
156. P. Vogel, J. Engel, Neutrino electromagnetic form factors. Phys. Rev. D **39**, 3378–3383 (1989)
157. Y. Wang et al., Improved limits on solar axions and bosonic dark matter from the CDEX-1B experiment using profile likelihood ratio method (2019). arXiv:1911.03085
158. S. Weinberg, A new light boson? Phys. Rev. Lett. **40**, 223–226 (1978)
159. F. Wilczek, Problem of strong *P* and *T* invariance in the presence of instantons. Phys. Rev. Lett. **40**, 279–282 (1978)
160. S.S. Wilks, The large-sample distribution of the likelihood ratio for testing composite hypotheses. Ann. Math. Stat. **9**(1), 60–62 (1938)
161. C. Wittweg, B. Lenardo, A. Fieguth, C. Weinheimer, Detection prospects for the second-order weak decays of ^{124}Xe in multi-tonne xenon time projection chambers (2020). arXiv:2002.04239
162. S.-C. Wu, Nuclear data sheets for A = 214. Nucl. Data Sheets **110**(3), 681–748 (2009)
163. J. Ye, Searches for WIMPs and axions with the XENON1T experiment. PhD thesis, UC San Diego, 2020
164. C. Zhang, D.-M. Mei, V.A. Kudryavtsev, S. Fiorucci, Cosmogenic activation of materials used in rare event search experiments. Astropart. Phys. **84**, 62–69 (2016)
165. A.P. Zhitnitskii, Possible suppression of axion-hadron interactions. Sov. J. Nucl. Phys. **31**, 260 (1980)
166. H. Zhong et al., Dark matter direct search sensitivity of the PandaX-4T experiment. Sci. China Phys. Mech. Astron. **62**(3), 31011 (2018)